HUMAN
GENE
THERAPY

This publication is based on presentations at the annual meeting of the Institute of Medicine and a workshop, cosponsored by the Institute and the National Academy of Sciences, held in Washington, D.C., on October 15–16, 1986. The views expressed are those of the participants and do not necessarily reflect those of the Institute and the Academy.

The Institute of Medicine was chartered by the National Academy of Sciences to enlist distinguished members of appropriate professions in the examination of policy matters pertaining to the health of the public. In this, the Institute acts under both the Academy's congressional charter responsibility to be an adviser to the federal government and its own initiative in identifying issues of medical care, research, and education.

The National Academy of Sciences was established in 1863 by Act of Congress as a private, nonprofit, self-governing membership corporation for the furtherance of science and technology, required to advise the federal government upon request within its fields of competence. Under its corporate charter the Academy established the National Research Council in 1916, the National Academy of Engineering in 1964, and the Institute of Medicine in 1970.

The National Research Council has become the principal operating agency of both the National Academy of Sciences and the National Academy of Engineering in the conduct of their services to the government, the public, and the scientific and engineering communities. It is administered jointly by both Academies and the Institute of Medicine.

HUMAN GENE THERAPY

Institute of Medicine
National Academy of Sciences

Eve K. Nichols

HARVARD UNIVERSITY PRESS
Cambridge, Massachusetts
London, England ▪

Library of Congress Cataloging in Publication Data

Nichols, Eve K.
   Human gene therapy.

   Drawn from the Annual Meeting of the Institute of Medicine, held
in Washington, D.C., Oct. 15–16, 1986.
   Includes bibliographies and index.
   1. Gene therapy. 2. Gene therapy—Moral and ethical aspects.
3. Gene therapy—Government policy—United States. I. Institute of
Medicine (U.S.). Meeting (1986 : Washington, D.C.) II. Title.
[DNLM: 1. Genetic Intervention. 2. Hereditary Diseases—therapy.
QZ 50 N617h]
RB155.8.N53   1988   616'.042   88-574
ISBN 0-674-41470-5 (alk. paper)
ISBN 0-674-41480-2 (pbk. : alk. paper)

Preface

Genetic diseases, though individually rare, when taken as a group have a large impact, affecting about five percent of infants in the United States. These disorders often cause early death or major disability, including mental retardation, pain, and crippling. In the vast majority of cases, effective treatment is unknown. It is in this context that the families and physicians of victims of genetic disorders, and researchers interested in the several thousand known genetic diseases, have seen new hope in the dazzling power of molecular genetics and recombinant DNA technology—in genetic engineering, in short.

But others see a darker side. They have raised questions about the impact genetic engineering of humans may have on the value we place on human life in all its diversity; they fear misuse of this powerful new technology. Their concerns may be based on religious or ethical grounds, or they may derive from the abuses of eugenics movements in the past.

It is likely that both the extreme hopes and the strongest fears about genetic engineering applied to humans stem from misconceptions about the nature of the relevant research and how far it has progressed. Accordingly, the Councils of the National Academy of Sciences and the Institute of Medicine convened a meeting of experts to provide a clear exposition of the scientific state-of-the-art, the technical hurdles that remain, and the public policy process that will precede any attempts to treat genetic disorders by genetic engineering.

This book is based on that meeting—"Human Somatic Cell Gene Therapy: Prospects for Treating Inherited Dis-

eases," held during the October 1986 annual meeting of the
Institute of Medicine—and a workshop that followed. It details new technologies that allow earlier and faster diagnosis
of some genetic diseases, and it reports the successes and the
disappointments of "traditional" approaches to treating these
diseases. It indicates the rapid pace at which new disease-causing genes have been identified and methods developed to
insert their normal counterparts into cells and animals. But it
also points out how difficult it has been to achieve adequate
functioning of the inserted genes (in studies with animals).
The book also highlights some of the social and ethical concerns raised by different types of genetic engineering and outlines the public policy process that is in place to address these
concerns in addition to the usual clinical questions about
safety and efficacy.

The information and views presented in this book are
based on the presentations at the meeting and workshop—the
speakers' contributions are acknowledged at the end of each
chapter. The special contribution of the author, Eve K.
Nichols, and the project coordinator, Barbara Filner, has been
to make the technical information accessible to a wider audience, one not trained in the language and methods of molecular biology and medical genetics. I hope that this book,
by clarifying what gene therapy might do—and what it cannot
do—will help our country, and affected individuals and their
families, make well-informed decisions about when and how
to apply the knowledge that this exciting research brings.

Frank Press
President
National Academy of Sciences

Contents

HUMAN
GENE
THERAPY

1 | The Potential of Gene Therapy

The baby's cesarean birth—on December 10, 2007—had been scheduled for weeks. The operating room team was fully scrubbed, and the special-care nurses in the isolation unit were ready.

When the infant appeared, the father commented on how healthy he looked. But everyone present knew that the howling newborn, already on his way to the brightly lit isolation room upstairs, had a rare, life-threatening genetic disease, severe combined immune deficiency (SCID). Prenatal diagnosis had shown that his body could not produce an essential protein, an enzyme called adenosine deaminase (ADA).[1] Without this enzyme, his immune system would fail to develop. He would be defenseless against even the most benign viruses, bacteria, and fungi.

The parents' previous child also had been born with SCID. Specialists in pediatric immunology had diagnosed his condition at the age of 14 months, but by then he was too ill to tolerate experimental medical procedures that might have prolonged his life. He died before his second birthday, worn out from intractable diarrhea and countless respiratory infections.

The outlook for the new baby was considerably brighter. Because the family's physicians had known in advance that he would be born with SCID, they had made arrangements to provide him with a germ-free environment—a closed world similar to that of the Houston SCID patient who became known nationwide as "the bubble boy." For the new baby, though, the bubble would be only a temporary home.

Like the bubble boy, the new baby lacked a genetically matched sibling. A bone marrow transplant from a matched donor would have had a high probability (80 percent or better) of repairing the baby's immune system, but the likelihood of finding such a donor outside the family is 1 in 100,000. Unmatched transplants are possible in SCID patients, but they are much less successful than matched transplants and the children often must be prepared with chemotherapy (which destroys any residual immunity, can cause sterility, and may have other harmful side effects).

Instead, the parents and physicians chose a different route. After a series of tests to determine the exact level of the child's existing immune function, the medical team removed a small amount of bone marrow from his hip. Cells from the bone marrow were grown in the laboratory and then infected with a special, modified virus. The virus had three important characteristics: it carried the genetic information (a single human gene) necessary to make the missing ADA protein; it could insert a single copy of this genetic information into the hereditary code of a human bone marrow cell; and it did not have the capability to reproduce or to cause disease—it served merely as a delivery system, or vector.

Two days later, the baby became his own bone marrow donor. Descendants of the bone marrow cells taken from his hip, now carrying the ADA gene, were injected into his bloodstream. Would the cells find their way back to the bone marrow and begin producing healthy immune cells?

Within a month, extensive blood tests showed that the autologous transplant had worked: normal lymphocytes—the white blood cells responsible for specific immunity—were appearing in greater numbers each day. At three months of age, the baby left the bubble in his mother's arms, ready to live in the outside world.

This brief, imaginary scene outlines the basic strategies that will be employed sometime in the future in the new field of human somatic cell gene therapy. (The term **somatic** refers to all cells in the body except the reproductive cells.) A child with a life-threatening genetic disease, caused by a known defect in a single gene, will be treated with the gene's normal counterpart. The normal gene, produced by recombinant DNA technology, will be inserted into a specific tissue in the body and will not be passed on to future generations. The goals will be the same as those associated with any form of medical treatment: reduction of human suffering and restoration of health.

Five years ago, many researchers believed that the first human trials of the procedure described above would occur before the end of 1986. But the tasks of delivering normal genes to recipient cells and of achieving appropriate expression of those genes (getting them to produce enough of their product to make a clinical difference) have proved much more difficult than expected. In addition, some techniques that appeared promising in cell culture have failed in animal studies, indicating the need for greater understanding of gene regulation inside the body.

The Impact of Genetic Diseases

Researchers in more than a dozen laboratories in the United States continue to pursue new techniques in gene therapy because of the promise the field holds for reducing the burden of genetic diseases. As a group, genetic disorders affect about 5 percent of liveborn infants in the United States.

Variations in DNA

Genetic diseases result from variations in the material that determines heredity, **deoxyribonucleic acid (DNA)**. DNA carries the information necessary for the maintenance and replication of cells. During cell division (see Figure 1.1), human DNA is visible under the microscope as 23 pairs of rod-like structures called **chromosomes** (one pair of sex chromosomes and 22 pairs of autosomal chromosomes; see Figure 1.2). Each chromosome contains thousands of **genes,** which are the basic functional units that code for specific proteins (either enzymes, which regulate chemical reactions in the cell, or essential chemical building blocks).

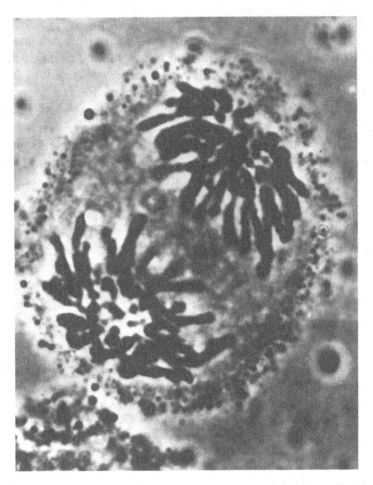

FIGURE 1.1 A human cell about to divide into two daughter cells. The cell contains a complete set of threadlike chromosomes for each daughter cell. SOURCE: March of Dimes Birth Defects Foundation.

Many genetic diseases are apparent at birth, but others may not appear until early childhood or later. Huntington disease, for example, usually does not become apparent until after age 30. Furthermore, many structural and biochemical defects present at birth are not the result of genetic disorders. Some infections (including rubella, cytomegalovirus, toxoplasmosis, untreated syphilis, gonorrhea, genital herpes, and

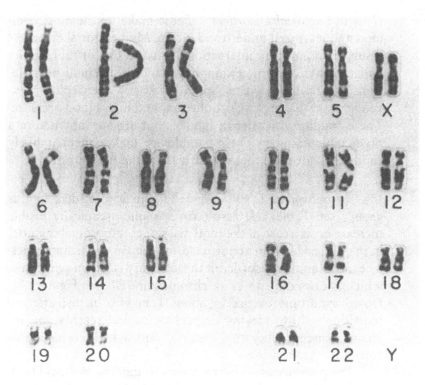

FIGURE 1.2 The systematic display of chromosomes from a single somatic cell is called a karyotype. Technicians select a cell in which the individual chromosomes are clearly visible. They photograph the cell through a microscope and then cut the chromosomes out of the photograph and pair them. By international convention, the human chromosome pairs are numbered and arranged in descending order of length. (The chromosomes are stained to reveal patterns of bands that aid in their identification.) The normal human female karyotype shown above has two X chromosomes and no Y chromosome. SOURCE: March of Dimes Birth Defects Foundation.

AIDS), as well as alcohol and other drugs and certain environmental factors, can cause a wide range of abnormalities in newborns. Specialists in genetic counseling can help parents determine the cause of a particular birth defect and the likelihood of its recurrence in future offspring.

To understand the potential impact of gene therapy on genetic disorders, it is helpful to divide them into three groups.

(1) Multifactorial Disorders. These make up the most common and least well understood group. Multifactorial disorders result from complex interactions between one or more genes and the environment. Examples include congenital heart lesions, neural tube defects (spina bifida and anencephaly), juvenile onset diabetes, schizophrenia, and high blood pressure. These conditions cluster in families but are not inherited in a predictable fashion. Their complexity makes them unlikely candidates for gene therapy, at least in the near future.

(2) Chromosomal Disorders. The genetic disorders in this group (see Table 1.1) result from a microscopically visible increase or decrease in the total mass of chromosomal material in the cell, or from an abnormal arrangement of chromosomes. The most familiar disorder in this category is Down syndrome, which is caused by an extra chromosome 21 (see Figure 1.3). Down syndrome occurs in about 1 in 600 births; affected individuals show varying degrees of mental retardation, increased susceptibility to infection, and a high incidence of congenital heart disease.

The presence of an extra chromosome (the technical term

TABLE 1.1. Prevalence of some common chromosomal disorders among live-born infants.

Disorder	Prevalence
Autosomal abnormalities	
Trisomy 21 (Down syndrome)	1 in 600
Trisomy 18	1 in 5,000
Trisomy 13	1 in 15,000
Sex chromosome abnormalities[a]	
Klinefelter syndrome	1 in 450 males
XYY syndrome	1 in 1,000 males
Triple-X syndrome	1 in 1,000 females
Turner syndrome	1 in 1,500 females

a. The sex chromosome abnormalities result from an abnormal number of sex chromosomes. For example, patients with Klinefelter syndrome usually have three sex chromosomes, two X and one Y. Individuals with Turner syndrome usually have only one sex chromosome, an X chromosome.

SOURCE: Adapted, with permission, from Stanbury et al., *The Metabolic Basis of Inherited Disease,* 5th ed. (New York: McGraw-Hill, 1983), table 1-3.

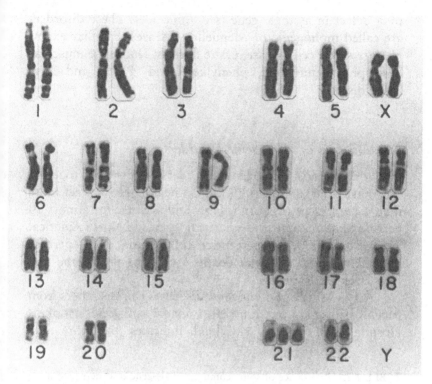

FIGURE 1.3   The karyotype of a girl with Down syndrome has an extra chromosome number 21 (three rather than the normal two). SOURCE: March of Dimes Birth Defects Foundation.

is **trisomy**) has a profound effect on development and metabolism. Excess gene products disrupt essential biochemical reactions throughout the body. Trisomy 18 and trisomy 13 (in which the extra chromosomes are number 18 and 13, respectively) also occur, but affected children have multiple congenital malformations and rarely survive infancy. Although gene therapy eventually may be able to ameliorate some of the symptoms of some chromosomal disorders, it probably will never be capable of curing any of these conditions.

(3) Single-Gene Disorders.   Researchers in gene therapy are focusing primarily on the third group of genetic disorders, which consists of almost 4,000 diseases known to be caused

by a defect in a single gene (see Table 1.2). These disorders
are called **monogenic** or Mendelian diseases. Familiar examples are sickle cell disease, cystic fibrosis, Duchenne muscular
dystrophy, hemophilia, phenylketonuria (PKU), and Tay-
Sachs disease.

Consequences of Monogenic Disorders

Individually, most monogenic disorders are rare, but together
they have a large impact on human health. They affect more
than 1 percent of liveborn infants and account for almost 10
percent of admissions to pediatric hospitals in North America.
They cause about 8.5 percent of child deaths and 7 percent
of stillbirths and neonatal deaths (occurring during the first
four weeks after birth).

A 1985 study of monogenic disorders by researchers from
McGill University in Montreal, Canada, and Johns Hopkins
Hospital in Baltimore, Maryland, indicates that more than

TABLE 1.2. Prevalence of some common single-gene disorders among
live-born infants.

Disorder	Estimated prevalence
Sickle cell disease	1 in 625 (U.S. blacks)
Cystic fibrosis	1 in 2,000 (U.S. whites)
Huntington disease	1 in 2,500
Duchenne muscular dystrophy	1 in 7,000
Hemophilia	1 in 10,000
Phenylketonuria (PKU)	1 in 12,000
Mucopolysaccharidoses[a] (all types together)	1 in 25,000
Glycogen storage diseases[b] (all types together)	1 in 50,000
Lesch-Nyhan syndrome	1 in 100,000

a. The mucopolysaccharidoses are lysosomal storage diseases (see Chapter 2). Examples include Hurler syndrome, Hunter syndrome, and Sanfilippo syndrome.

b. Examples of glycogen storage diseases include von Gierke disease, Pompe disease, and Hers disease.

SOURCE: Adapted, with permission, from Stanbury et al., *The Metabolic Basis of
Inherited Disease,* 5th ed. (New York: McGraw-Hill, 1983), table 1-4.

half of all known monogenic disorders lead to early death. Three-quarters of those compatible with life beyond infancy limit access to schooling, ability to work, or both. Two-thirds impair the reproductive capability of affected individuals.

In a companion study, the Canadian and American scientists showed that modern medicine and surgery have relatively little to offer most patients with single-gene disorders. In the 351 single-gene diseases they examined, the researchers found that treatment increased lifespan to normal in only 15 percent of the disorders, increased reproductive capability in 11 percent, and increased social adaptation in 6 percent.

The researchers conducted a separate analysis of the effect of treatment on the best understood single-gene diseases, inborn errors of metabolism. These are diseases in which the absence of an enzyme, or the presence of a defective enzyme, produces a disturbance in a specific biochemical reaction (adenosine deaminase deficiency is a good example). They concluded that a few patients benefit tremendously from existing dietary, pharmacologic, and surgical interventions, but that most benefit very little. Treatment produced complete relief in only 12 percent of the 65 inherited metabolic disorders included in the sample and a partial response in 40 percent. It had no effect on the remaining 48 percent of diseases.

These figures illustrate the immense collective burden associated with genetic diseases. The extent to which somatic cell gene therapy will be able to reduce this burden is not yet clear. Initially, the technique will be useful only for a handful of monogenic disorders; but as researchers learn more about specific gene defects, about the introduction of genes into human cells, and about the regulation of inserted genes, it could become a standard part of medical practice.

Ethical and Social Concerns

In the early 1970s, when genetic engineering first became possible, many lay persons and scientists expressed strong concerns about its potential social and ethical implications. Primary among these concerns was the fear that genetic engineering would be used to "remake" the human race. Images arose of Frankenstein-like monsters with superhuman intellectual and physical powers, or, alternatively, of workers genetically engineered to be docile and subservient.

Through extensive public discussions, involving civic and religious leaders, biomedical researchers, ethicists, federal policymakers, and others, a framework was developed to help put these concerns in perspective. Genetic engineering, as it relates to the insertion of genes into human cells, can be divided into four distinct categories:

- somatic cell gene therapy
- germ line gene therapy
- enhancement genetic engineering
- eugenic genetic engineering.

Somatic Cell Gene Therapy

Somatic cell gene therapy is the only technique now considered appropriate for use in human beings. The insertion of a single gene into the somatic cells of an individual with a life-threatening genetic disease is intended solely to eliminate the clinical consequences of the disease; the inserted gene is not passed on to future generations. This kind of therapy is much different from the other three forms of genetic engineering, which pose different technical problems as well as major ethical concerns about experimentation with human embryos and the long-term effects of changing the gene pool of the species.

Germ Line Gene Therapy

Germ line gene therapy, which has been performed successfully in several animal studies, involves the insertion of a healthy gene into the fertilized egg of an animal that has a specific genetic defect. Every cell in the body, including the reproductive cells, acquires the new gene. For example, scientists at Columbia University recently corrected a hereditary blood disorder (beta thalassemia) in mice by inserting copies of a normal blood protein gene into fertilized mouse eggs. The treated eggs were transferred to foster mothers, and several developed into healthy mice with blood cells that looked and functioned normally. The normal genes were passed on to later generations when the original mice were mated.

Three overwhelming technical problems prevent consideration of this technique for use in human beings. The first is that scientists have no way of diagnosing genetic disorders in the fertilized egg. Depending on the mode of inheritance (see Chapter 2), each offspring in a family affected by a monogenic disorder has either a one-in-two or a one-in-four chance of developing the disease; statistically, at least half of all egg–sperm unions could result in a healthy embryo. It would not be rational or ethical to subject all fertilized eggs to the risks of an experimental procedure when the potential benefits would be limited to the embryos carrying the abnormal gene.

The second problem is that the procedure most often used to insert genes into fertilized eggs—injection with a microscopically guided glass needle—has a high failure rate. The vast majority of eggs are so damaged by the microinjection process that they do not develop into live offspring. In a 1983 study by Ralph L. Brinster of the University of Pennsylvania and his coworkers, only 11 of 300 mouse eggs injected with an immunoglobulin gene resulted in live births, and only 6 animals (2 percent) carried the new gene.

The remaining problem is lack of control over where the gene is inserted into the embryo's genetic machinery. This issue also arises in somatic cell gene therapy, but when every cell in the body is affected the consequences of random insertion of a new piece of DNA appear to be much greater. In several cases, genes have been expressed in inappropriate tissues. For example, researchers have detected hemoglobin production in the muscles and testes of some laboratory animals that received a hemoglobin gene shortly after conception. In other animal studies, the disruption of normal genes by microinjected DNA has resulted in fetal deaths caused by the absence of an essential gene product.

Enhancement Genetic Engineering

The disparity between potential risks and benefits is even more striking for enhancement genetic engineering, because the goal would no longer be treatment of a genetic disease. Enhancement would involve the use of gene therapy to alter a specific characteristic in a preferred direction. The example cited most often is the insertion of a growth hormone gene to increase stature in a healthy child. The risks associated with such a procedure could be numerous. Scientists know very little about the complex feedback mechanisms that control essential biochemical reactions in the body. The imbalance created by excessive quantities of one gene product could have many unanticipated effects. Besides, the same "benefits" could be achieved by the administration of growth hormone itself. Although the direct use of growth hormone in healthy children also has disturbing ethical and medical implications, at least the physician retains some control over the outcome. Drug treatments can be stopped; insertion of an extra gene would not be reversible.

Eugenic Genetic Engineering

Eugenic genetic engineering refers to the use of recombinant DNA technology to alter traits such as intelligence, personality, and organ formation. These traits are controlled by hundreds—perhaps thousands—of genes interacting with multiple factors in the environment. The genes involved have not been identified. Furthermore, scientists do not have the tools to manipulate these complex traits, and do not expect to have them in the foreseeable future.

Fears that eugenic genetic engineering might be used by an amoral government to implement coercive social programs have fueled much of the anxiety about the development of gene therapy. These fears simply are not realistic in terms of today's scientific achievements. Public discussion about potential abuses of new technologies is an important part of the decision-making process in modern society. However, discussions about eugenic genetic engineering should not be viewed as an indication that the technique is either imminent or a likely outcome of current research. Indeed, eugenic genetic engineering may never be possible because of the extreme complexity of the systems involved.

Public Review

Careful analyses of the medical and social implications of gene therapy by the President's Commission for the Study of Ethical Problems in Medicine and Biomedical and Behavioral Research and by the congressional Office of Technology Assessment (OTA) have helped allay the public's fears about the misuse of the new technology. The 1984 OTA report, *Human Gene Therapy—A Background Paper*, concludes:

> Civic, religious, scientific, and medical groups have all accepted, in principle, the appropriateness of gene therapy of somatic cells in humans for specific genetic diseases ... Whether somatic cell gene therapy will become a practical medical technology will thus depend on its safety and efficacy, and the major question is when to begin clinical trials, not whether to begin them at all. The quality that distinguishes somatic cell gene therapy most strongly from other medical technologies is not technical, but rather the public attention that is likely to attend its commencement.

Existing Review Mechanisms

The first clinical trials will be scrutinized both by local and national review boards according to a procedure established by the Recombinant DNA Advisory Committee of the National Institutes of Health. Researchers will be required to answer extensive questions outlined in a document titled, "Points to Consider in the Design and Submission of Human Somatic-Cell Gene Therapy Protocols." In addition, the U.S. Food and Drug Administration (FDA) will review the proposed trials under its mandate to monitor the use of investigational drugs and biologics.

Ethicist LeRoy Walters speculates that if somatic cell gene therapy proves to be both safe and effective, the review process gradually will be decentralized, with a concomitant trend toward deregulation. At that point, somatic cell gene therapy trials will have to meet the same criteria as trials of any new medical or surgical technique: review committees will seek evidence that the probable benefits to the patient, established through appropriate laboratory studies, outweigh the probable risks.

Both the President's Commission and the OTA have emphasized that the consensus about the potential role of somatic cell gene therapy does not extend to germ line gene therapy or to genetic manipulations designed to alter characteristics in healthy individuals. Extensive public discussion of pertinent safety and ethical issues will be required to determine whether these other forms of genetic engineering will be acceptable in our society.

Contributions to Other Areas of Science

Even as we wait for clinical trials to begin, basic research on the control of genetic diseases is providing new insights into the intricate workings of the human body. Scientists exploring the potential of gene therapy have made major advances in understanding the development of the normal immune system, the mechanisms that switch genes on and off, and the interactions between viruses and mammalian cells. Much of this knowledge already has been integrated into other areas of research, including cancer therapy. More than 300 years ago, the great British physician William Harvey described the potential benefits of studying variations in the human population:

> Nature is nowhere accustomed more openly to display her secret mysteries than in cases where she shows tracings of her workings apart from the beaten paths. Nor is there any better way to advance the proper practice of medicine than to give our minds to the discovery of the usual law of nature by careful investigation of cases of rare forms of disease.

—— ACKNOWLEDGMENTS ——

Chapter 1 is based on the presentations of W. French Anderson, Rochelle Hirschhorn, David W. Martin, Jr., Leon E. Rosenberg, LeRoy Walters, and James B. Wyngaarden.

—— NOTES ——

1. The worldwide incidence of SCID is estimated to be between 1 in 100,000 and 1 in 500,000 births. About one-fifth of SCID cases result from ADA deficiency.

—— SUGGESTED READINGS ——

Anderson, W. French. 1984. "Prospects for Human Gene Therapy." *Science* 226:401–409.

———. 1985. "Human Gene Therapy: Scientific and Ethical Considerations." *Journal of Medicine and Philosophy* 10:275–291.

Brinster, Ralph L., Kindred A. Ritchie, Robert E. Hammer, Rebecca L. O'Brien, Benjamin Arp, and Ursula Storb. 1983. "Expression of a Microinjected Immunoglobulin Gene in the Spleen of Transgenic Mice." *Nature* 306:332–336.

Costa, Teresa, Charles R. Scriver, and Barton Childs. 1985. "The Effect of Mendelian Disease on Human Health: A Measurement." *American Journal of Medical Genetics* 21:231–242.

Costantini, Frank, Kiran Chada, and Jeanne Magram. 1986. "Correction of Murine β-Thalassemia by Gene Transfer into the Germ Line." *Science* 233:1192–1194.

Hayes, Ailish, Teresa Costa, Charles R. Scriver, and Barton Childs. 1985. "The Effect of Mendelian Disease on Human Health: II. Response to Treatment." *American Journal of Medical Genetics* 21:243–255.

Human Gene Therapy—A Background Paper. 1984. Washington, D.C.: U.S. Congress, Office of Technology Assessment. OTA-BP-BA-32.

The New Human Genetics: How Gene Splicing Helps Researchers Fight Inherited Disease. 1984. Written by Maya Pines. Bethesda, MD: National Institute of General Medical Sciences, National Institutes of Health. NIH Publication no. 84-662.

President's Commission for the Study of Ethical Problems in Medicine and Biomedical and Behavioral Research. 1982. *Splicing Life: The Social and Ethical Issues of Genetic Engineering with Human Beings*. Washington, D.C.: U.S. Government Printing Office. Stock no. 83-600500.

Rosenberg, Leon E. 1985. "Can We Cure Genetic Disorders?" In Aubrey Milunsky and George J. Annas, eds., *Genetics and the Law III*, pp. 5–13. New York: Plenum Press.

Stanbury, John B., James B. Wyngaarden, Donald S. Fredrickson, Joseph L. Goldstein, and Michael S. Brown. 1983. *The Metabolic Basis of Inherited Disease*. 5th edition. New York: McGraw-Hill.

2 | Choosing Disease Candidates

Scientists asked to imagine the ideal gene therapy of the future envision a situation in which one injection would be enough to cure a genetic disease. The injection would consist of multiple copies of a single normal gene. Each gene would be packaged for direct delivery to a particular type of cell—for example, bone marrow cells in sickle cell disease or muscle cells in Duchenne muscular dystrophy. Once inside a cell, the gene would replace its abnormal counterpart and all signs of disease would disappear.

Many technological hurdles stand between this ideal and the current state of the art in gene therapy. The first major hurdle is the need for more information about the basic defects responsible for most genetic diseases. Medical researchers have identified almost 4,000 single-gene disorders; they have a clear understanding of the disease process in about 250.

Another major hurdle is that scientists do not have the controls necessary to send genetic material into the body and ensure that it will end up inside a particular cell, at a specific location in the cell's genetic machinery. If they attempted such a feat with today's technology, most or all of the genetic material would be destroyed by chemicals in the blood that degrade naked DNA or by the immune response to foreign materials. Any incorporation of the new gene into cells would occur randomly; even reproductive cells might be affected.

Current gene therapy techniques require that insertion of a healthy gene take place outside the body. This implies the capability to remove cells from an affected patient, treat them in the laboratory, and give them back to the patient in a manner that is consistent with continued growth and development. At present, only two types of cells can be manipulated in this way: bone marrow cells and skin cells. Early gene therapy trials probably will employ bone marrow cells, in part because of the vast experience with bone marrow transplantation.

Technological limitations and the need for extreme caution in the application of any new approach to health care shape the criteria for selection of disease candidates for gene therapy. For each candidate, clinical and laboratory evidence must be available to show that

(1) the disease is life-threatening and incurable without gene therapy;

(2) the organ, tissue, and cell types affected by the disease have been identified;

(3) the normal counterpart of the defective gene has been isolated and cloned;

(4) the normal gene can be introduced into a substantial fraction of cells from the affected tissue; or that introduction of the gene into an available target tissue, such as bone marrow, will somehow alter the disease process in the tissue affected by the disease;

(5) the gene can be expressed adequately (it will direct the production of enough normal protein to make a difference);

(6) techniques are available to verify the safety of the procedure.

The remainder of this chapter focuses on the first two criteria. Scientists must understand disease candidates for gene therapy at many different levels. Every genetic disorder results from a primary defect in DNA structure, a **mutation**. Harmful mutations produce a disturbance in protein structure and function, which in turn leads to a disruption in cell and organ function. (Not all mutations are harmful—some have no effect on protein function and a few actually improve the survival potential of the affected individual.)

As explained in Chapter 1, gene therapy will be appropriate only for genetic disorders caused by a mutation in a single gene. Researchers can identify these diseases by studying inheritance patterns in affected families. Single-gene disorders show one of three common patterns of inheritance: (1) autosomal recessive; (2) autosomal dominant; or (3) X-linked.

Inheritance Patterns

Autosomal recessive and autosomal dominant patterns involve genes on the non-sex, or **autosomal,** chromosomes. Human beings have 44 autosomal chromosomes—22 derived from the mother and 22 derived from the father. Each maternal chromosome has a matching, or homologous, paternal chromosome; thus, each person has two genes for each autosomal trait. If the two genes that code for a particular trait are identical, the individual is said to be **homozygous** for that trait; if the genes are different, the individual is **heterozygous.**

Autosomal Recessive Disorders

Recessive diseases require the presence of a pair of mutant genes (see Figure 2.1). Parents of an affected child usually are healthy carriers; they each have one normal gene and one defective gene. Each of their offspring has a 25 percent chance of inheriting two mutant genes (and therefore developing the genetic disease); a 25 percent chance of inheriting two normal genes; and a 50 percent chance of inheriting one normal gene and one mutant gene (which would make the child a carrier like the parents). Autosomal recessive disorders affect males and females in equal proportions.

Familiar autosomal recessive disorders include sickle cell disease, cystic fibrosis, Tay-Sachs disease, and phenylketonuria, or PKU. Scientists estimate that every human being inherits six or seven recessive mutations that, if transmitted to offspring who happen to inherit the same recessive mutation from the other parent, could cause serious illness or death. Recessive disorders usually are diagnosed in infancy or early childhood.

AUTOSOMAL RECESSIVE INHERITANCE

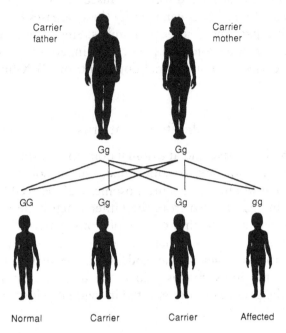

FIGURE 2.1 Autosomal recessive disease requires the presence of two mutant genes in the affected individual. Both parents carry a defective gene (**g**) and a normal gene (**G**)—in most cases, the single normal gene is sufficient for normal function and the parents are healthy. Each child has a 25 percent risk of inheriting a "double dose" of the **g** gene, which may cause a serious genetic defect; a 25 percent chance of inheriting two normal genes; and a 50 percent chance of being a carrier like the parents. SOURCE: Adapted from *The New Human Genetics: How Gene Splicing Helps Researchers Fight Inherited Disease*, written by Maya Pines (Bethesda, MD: National Institute of General Medical Sciences, 1984), NIH Publication no. 84-662.

Autosomal Dominant Disorders

In autosomal dominant disorders, an abnormal gene "dominates" its normal counterpart. Thus, one mutant gene is sufficient to cause disease (see Figure 2.2). Each child of a parent with an autosomal dominant disorder has a 50 percent chance of inheriting the mutant gene and, therefore, of developing the disease. (If both parents are affected—a very rare occurrence—each offspring has a 75 percent chance of inheriting

AUTOSOMAL DOMINANT INHERITANCE

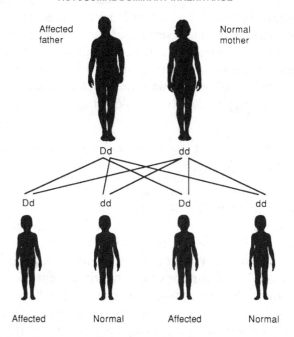

FIGURE 2.2 In autosomal dominant disorders, one parent has a single faulty gene (D), which dominates its normal counterpart (d). Each child has a 50 percent risk of inheriting the D—and the disorder—from the affected parent. SOURCE: Adapted from *The New Human Genetics: How Gene Splicing Helps Researchers Fight Inherited Disease*, written by Maya Pines (Bethesda, MD: National Institute of General Medical Sciences, 1984), NIH Publication no. 84-662.

the mutant gene.) Males and females are affected in equal proportions. Normal offspring of an affected individual do not have the abnormal gene and cannot transmit it to their children.

Examples of autosomal dominant diseases include Huntington disease, familial hypercholesterolemia (in which an abnormal increase in the amount of cholesterol in the blood leads to early coronary artery disease), and polycystic kidney disease (a major cause of kidney failure).

As a group, autosomal dominant disorders have two characteristics that distinguish them from recessive syndromes:

(1) symptoms may not appear until adolescence or later, and (2) the clinical picture often varies from one person to another. In some cases, the clinical signs may be so diverse that physicians fail to recognize that affected family members have the same genetic abnormality. For example, in multiple endocrine neoplasia syndrome, the same genetic defect may cause stomach ulcers, kidney stones, benign skin tumors, or vision problems.

X-Linked Disorders

X-linked diseases affect males and females differently because the genes responsible are located on one of the two sex chromosomes, the X chromosome. A girl has two X chromosomes, one inherited from each parent. A boy has only one X chromosome, and it always comes from his mother; his other sex chromosome is a Y chromosome, inherited from his father.

In typical X-linked disorders, the mother has a mutant gene on one of her X chromosomes (see Figure 2.3). Each of her children has a 50 percent chance of inheriting the abnormal gene, but only her sons develop the disease. Daughters who inherit the defective gene usually are free of symptoms because the normal gene on their second X chromosome (inherited from their fathers) produces enough normal protein to compensate for the defect; however, they can pass the disease on to their own sons. (This description assumes that the X-linked trait is recessive; X-linked dominant traits are rare and have a different pedigree pattern.)

Duchenne muscular dystrophy, hemophilia A, and Lesch-Nyhan syndrome are among the better known X-linked disorders. X-linked traits are never passed from father to son, because a man contributes his Y chromosome to his male offspring, never his X chromosome.

Autosomal and X-linked recessive disorders are more likely to be among the first candidates for gene therapy than autosomal dominant diseases. The most important reason is that insertion of a normal gene—the only form of gene therapy expected to be ready for clinical trials in the near future—

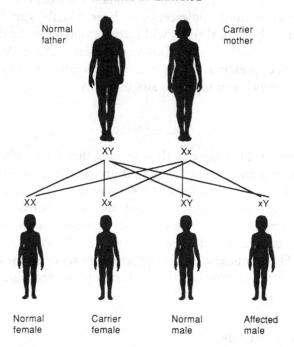

X-LINKED INHERITANCE

Normal father — Carrier mother

XY — Xx

XX / Xx / XY / xY

Normal female / Carrier female / Normal male / Affected male

FIGURE 2.3 In X-linked inheritance, the mother has a defective gene on one of her two sex chromosomes (**x**). She is protected against the associated disease because her normal sex chromosome (**X**) compensates for the defect. The father has normal male sex chromosomes (**X** and **Y**). Each male child has a 50 percent risk of inheriting the faulty **x** and the disorder, and a 50 percent chance of inheriting the normal **X** chromosome. Each female child has a 50 percent risk of inheriting the faulty **x** and becoming a carrier like her mother, and a 50 percent chance of inheriting two normal **X** chromosomes. SOURCE: Adapted from *The New Human Genetics: How Gene Splicing Helps Researchers Fight Inherited Disease*, written by Maya Pines (Bethesda, MD: National Institute of General Medical Sciences, 1984), NIH Publication no. 84-662.

may not be sufficient to correct an autosomal dominant disorder. Gene therapy for dominant disorders probably will require modification or inactivation of the abnormal gene responsible for the disease, a much more complex task than gene insertion. Also, benefit/risk assessments are more likely to be favorable for gene therapy if the target disease is lethal early in life; some autosomal dominant disorders do not become evident until adolescence or later. Finally, rapid evaluation of early clinical trials will be most reliable if the diseases involved present a fairly uniform clinical picture. As noted earlier, the presentation of many dominant diseases varies tremendously, even within a single family.

Gene Defects

The severity of a genetic disease in a given individual may depend to some extent on the nature of the gene defect responsible for the disease. Over the past 15 years, researchers have learned that many inherited disorders once believed to be uniform clinical entities actually represent a spectrum of diseases caused by different mutations in the same or related genes. The signficance of this observation will become evident after a brief review of the relationship between genes and proteins.

The Genetic Code

Genes are essentially pieces of DNA (see Figure 2.4). Each gene carries the instructions for a specific protein in a four-letter alphabet determined by the chemical structure of the DNA molecule. The "letters" in the alphabet are DNA subunits called **nucleotides** (see Figure 2.5). Each nucleotide consists of one of four nitrogen **bases**—adenine (A), thymine (T), guanine (G), or cytosine (C)—linked to a sugar molecule and a phosphate group.

The sequence of nucleotides in a gene determines its message, but accurate translation of a gene into an appropriate protein molecule requires that the letters be grouped into "words." In the 1960s, researchers at the National Institutes of Health, New York University, and the University of Wisconsin identified the basic "word" in the language of the genetic code. The unit is called a **codon**; it consists of three adjacent nucleotides (see Figure 2.6). Each codon corresponds to one of 20 **amino acids** (basic protein building blocks) or to a signal to start or stop the construction of an amino acid

DNA STRUCTURE

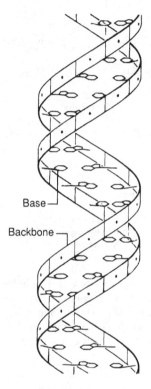

FIGURE 2.4 The DNA molecule is a long double-stranded chain. Each strand is made up of millions of minute subunits called nucleotides. A nucleotide is composed of three parts: a sugar, a phosphate group, and a flattened structure called a base. The sugar and phosphate group of each nucleotide contribute to the backbone of the DNA strand. The backbones of the two strands wind around each other to form a double helix. The bases, which are perpendicular to the sugars, tend to stack one on top of another, much like steps in a spiral staircase. The four bases in DNA are adenine (**A**), guanine (**G**), thymine (**T**), and cytosine (**C**). The information carried by DNA is coded in sequences of nucleotide bases. SOURCE: Reprinted with permission, from Karl Drlica, *Understanding DNA and Gene Cloning: A Guide for the Curious* (New York: John Wiley, 1984), figure 3-1. Copyright ©1984 by John Wiley & Sons, Inc.

THE FOUR NUCLEOTIDE BASES IN DNA

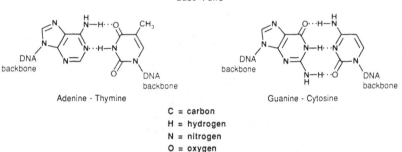

FIGURE 2.5 The two strands of DNA in the double helix fit together in a very precise way. Whenever an adenine occurs in one strand, a thymine must occur opposite it in the other strand. Similarly, guanine always pairs with cytosine. Hydrogen bonds between the complementary base pairs (**A** and **T, G** and **C**) hold the two strands of the double helix together.

chain. For example, the codon adenine-guanine-cytosine (AGC) codes for the amino acid serine. The codon TAA is one of three stop signals. There are 64 possible codons and only 20 amino acids, so more than one codon can translate into the same protein building block.

Types of Mutations

The cell's genetic machinery reads each gene in a linear fashion, always beginning at the same end. Mutations can disrupt the reading process, changing the structure of the corresponding protein, in several different ways.

TRANSLATING DNA INTO PROTEIN

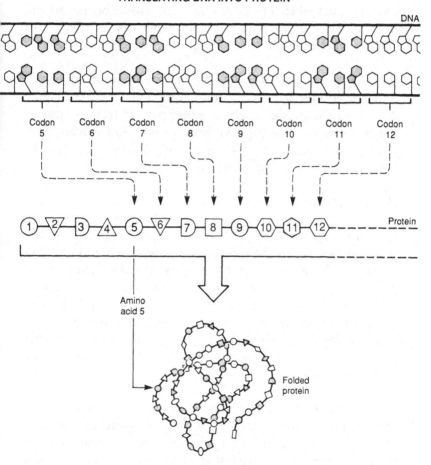

FIGURE 2.6 Information in DNA is arranged in a series of three-letter words called codons (the letters are individual nucleotides in the DNA molecule). Each codon specifies a particular amino acid. For example, codon 5 has information for the fifth amino acid in the protein. The overall shape and activity of a protein depend on the precise order of amino acids. The information for this order is stored in the DNA. SOURCE: Reprinted, with permission, from Karl Drlica, *Understanding DNA and Gene Cloning: A Guide for the Curious* (New York: John Wiley, 1984), figure 2-3. Copyright ©1984 by John Wiley & Sons, Inc.

Gross mutations.   Gross mutations involve the duplication, deletion, or translocation of multiple nucleotides. (Translocation occurs when a piece of one chromosome breaks off and fuses to another chromosome.) In some cases, stretches of DNA containing hundreds of genes may be deleted.

Frame-shift mutations.   The deletion or addition of one or two nucleotides is called a frame-shift mutation (see Figure 2.7). Such a mutation results in a new reading frame (regrouping of the nucleotide triplets). Every codon downstream from the mutation site changes—a completely different set of amino acids is incorporated into the protein.

Point mutations.   The simplest mutations are point mutations, in which one nucleotide is substituted for another. The effect of a point mutation depends on the nature of the substitution (see Figure 2.8). About 23 percent of random point mutations have no effect on protein structure, because the altered codon codes for the same amino acid as the original codon. For example, the DNA codons TTT and TTC both code for the amino acid phenylalanine, so a mutation that changed the final T in the TTT triplet to a C would not alter the protein product. This type of mutation is called a **synonymous mutation.**

Sickle cell disease is the best known example of the second type of point mutation, a **missense mutation.** Missense mutations result in the replacement of one amino acid for another in a protein chain. In 1956, Vernon Ingram, then of Cambridge University, showed that sickle cell hemoglobin was altered by the substitution of the amino acid valine for glutamic acid in the beta-globin subunit of the hemoglobin molecule. (Hemoglobin, the oxygen-carrying substance in the blood, is composed of four protein chains called **globins.** Most hemoglobin in adults consists of two alpha-globin chains and two beta-globin chains.) Later, researchers traced the substitution to a single nucleotide. The codon GAG in the normal beta-globin gene is changed to GTG in the sickle cell gene. GAG codes for glutamic acid, while GTG codes for valine.

The third type of point mutation is called a **nonsense**

FRAME-SHIFT MUTATION

(a) Normal DNA Molecule

DNA: —T—T—C—C—G—G—T—G—G—T—C—G—G—C—T—C—T—T—

Protein: — phe — arg — trp — ser — ala — leu —

(b) Mutation: G Deleted from Second Codon

DNA: —T—T—C—C—G—☒—T—G—G—T—C—G—G—C—T—C—T—T—

(c) Mutant DNA Molecule

DNA: —T—T—C—C—G—T—G—G—T—C—G—G—C—T—C—T—T—

Protein: — phe — arg — gly — arg — leu —

```
                           KEY
       DNA Codon              Amino Acid
         CGG              Arginine  (arg)
         CGT              Arginine  (arg)
         CTC              Leucine  (leu)
         CTT              Leucine  (leu)
         GCT              Alanine  (ala)
         GGT              Glycine  (gly)
         TCG              Serine  (ser)
         TGG              Tryptophan  (trp)
         TTC              Phenylalanine  (phe)
```

FIGURE 2.7 (a) A normal nucleotide sequence for one strand of DNA codes for a protein having the six amino acids shown below it. (b) Deletion of one base throws the reading frame out of register. (c) Every codon downstream from the mutation site changes, completely altering the structure of the resulting protein. SOURCE: Adapted, with permission, from Karl Drlica, *Understanding DNA and Gene Cloning: A Guide for the Curious* (New York: John Wiley, 1984), figure 4-6. Copyright ©1984 by John Wiley & Sons, Inc.

POINT MUTATIONS

(a) Normal Molecule

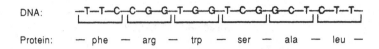

DNA: —T—T—C—C—G—G—T—G—G—T—C—G—G—C—T—C—T—T—

Protein: — phe — arg — trp — ser — ala — leu —

(b) Synonymous Mutation

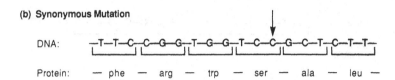

DNA: —T—T—C—C—G—G—T—G—G—T—C—C—G—C—T—C—T—T—

Protein: — phe — arg — trp — ser — ala — leu —

(c) Missense Mutation

DNA: —T—T—C—C—G—G—T—G—G—C—C—G—G—C—T—C—T—T—

Protein: — phe — arg — trp — pro — ala — leu —

(d) Nonsense Mutation

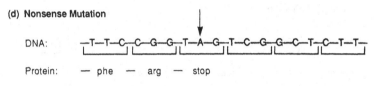

DNA: —T—T—C—C—G—G—T—A—G—T—C—G—G—C—T—C—T—T—

Protein: — phe — arg — stop

KEY	
DNA Codon	**Amino Acid**
CCG	Proline (pro)
CGG	Arginine (arg)
CTT	Leucine (leu)
GCT	Alanine (ala)
GGT	Glycine (gly)
TAG	Stop
TCC	Serine (ser)
TCG	Serine (ser)
TGG	Tryptophan (trp)
TTC	Phenylalanine (phe)

FIGURE 2.8 (a) A normal nucleotide sequence for one strand of DNA codes for a protein having the six amino acids shown below it. (b) The substitution of a C for the G in the fourth codon is called a synonymous mutation, because the protein molecule does not change. (c) If a C is substituted for the T in the fourth codon, the result is a missense mutation: the mutant protein has a proline where serine is normally located. (d) If the first G in the third codon is changed to an A, a nonsense (stop) codon is created and protein synthesis halts—the mutant protein is shorter than the normal protein. SOURCE: Adapted, with permission, from Karl Drlica, *Understanding DNA and Gene Cloning: A Guide for the Curious* (New York: John Wiley, 1984), figure 4-6. Copyright ©1984 by John Wiley & Sons, Inc.

mutation. Nonsense mutations change codons that normally code for a particular amino acid into "stop" codons. Protein production halts when the genetic machinery attempts to read the altered codon. Nonsense mutations have been found in many patients with thalassemia, a group of blood disorders caused by defective production of either alpha globin or beta globin. (Decreased production of one hemoglobin subunit leads to an abnormal build-up of the other, which disrupts the development of red blood cells.)

Thalassemia is an example of a very heterogeneous genetic disease. Children with severe thalassemia appear healthy at birth, but become pale and listless during the first year or two of life. Without treatment, they succumb to heart failure or infection at an early age. In contrast, children with mild thalassemia often lead completely normal lives. The differences in clinical outcome often can be traced to the nature of the mutation in the globin genes.

Gene Regulation

Research on thalassemia illustrates the importance of understanding gene regulation as well as gene structure. Five years ago, if someone had asked medical geneticists to name the disease most likely to be the first candidate for early gene therapy trials, many would have said beta thalassemia. It seemed like a perfect candidate. Researchers knew the identity of the mutant gene and had the technological capability to manipulate red blood cell precursors (bone marrow cells) outside the body. The disease was relatively common and, in its most severe form, usually lethal at an early age.

Subsequent laboratory studies revealed, however, that the development of gene therapy for beta thalassemia would be more difficult than expected. Scientists could introduce the normal beta-globin gene into an affected cell, but they could not control its expression (the amount of beta globin produced). Healthy cells balance globin production very precisely to prevent the build-up of one type of hemoglobin subunit or the other.[1]

Fortunately, the same level of precision is not required

for all gene products. Clinical evidence indicates, for example, that efforts to cure adenosine deaminase deficiency (see Chapter 1) with gene therapy probably would be successful if the level of enzyme inside immune cells could be raised to only 5 percent of normal. Moreover, an ADA level in blood serum that was 30 to 40 times normal probably would not be harmful to body tissues. This large margin of safety makes ADA deficiency an attractive candidate for early gene therapy trials.[2]

Defects in Protein Structure

With the exception of disorders affecting hemoglobin synthesis, most genetic diseases that have been characterized result from mutations in genes coding for enzymes. **Enzymes** regulate the speed of biochemical reactions in the body. Several different types of problems can occur when deficient or absent enzyme activity alters a crucial metabolic pathway.

Many of the classic inborn errors of metabolism result from an accumulation of the molecules on which enzymes act. Excess amounts of some of these molecules, called **substrates,** can be highly toxic. For example, in the disorder phenylketonuria (PKU), reduced activity of a specific enzyme (phenylalanine hydroxylase) leads to an abnormal accumulation of the amino acid phenylalanine and some of its biochemical derivatives. At high levels, these substances interfere with the normal development of brain cells, resulting in mental retardation. Other examples of diseases caused by the buildup of toxic substances include galactosemia (accumulation of galactose leads to vomiting, liver disease, cataracts, and mental retardation), maple syrup urine disease (accumulation of amino acids and ketoacids causes neonatal death or severe brain damage); and methylmalonic acidemia (accumulation of methylmalonate leads to developmental retardation).

In another large group of genetic disorders, called lysosomal storage diseases, inherited enzyme deficiencies disrupt the orderly renewal of cell constituents. Normal cells undergo a constant rebuilding process. Large molecules are broken down and the pieces are either reused or discharged into the extracellular space. The degradation process takes place in intracellular compartments called **lysosomes**; the different enzymes that degrade macromolecules are called lysosomal enzymes.

When one of the lysosomal enzymes is absent or defective, the chemical compound it acts on accumulates and cannot be removed from the cell. Lysosomes swell and eventually begin to interfere with other cell processes. The clinical signs of the lysosomal storage disorders depend on the nature of the stored material and on the function of the tissues involved.

Some of the more common diseases in this category are Tay-Sachs disease (substrate accumulates in the lysosomes of nerve cells, causing blindness, paralysis, and death by age five); Gaucher disease (substrate accumulates in large white blood cells in the bone marrow, spleen, and liver, causing bleeding disorders and enlargement of the affected organs; see Figure 2.9); Fabry disease (substrate accumulates in the lysosomes of blood vessels, leading to kidney failure and cardiovascular disease); and Hurler syndrome (substrate accumulates in many different tissues, resulting in skeletal abnormalities, spleen and liver enlargement, cataracts, and progressive mental retardation).

If an enzyme defect involves a metabolic reaction in which a complex molecule is constructed, rather than degraded, symptoms may be caused by deficiency of the end product. In one sense, scurvy could be considered an example of this type of genetic disease—one common to all human beings. Unlike other mammals, primates and guinea pigs lack an enzyme necessary to synthesize ascorbic acid, or vitamin C. Fortunately, we have access to many dietary sources of vitamin C, so the disease is rare. More conventional examples of diseases caused by missing end products include albinism, in which persons lack the pigment melanin, and familial goiter syndromes, in which affected individuals cannot produce the thyroid hormone thyroxine.

In a very few cases, genetic defects result in enzymes that are more active than their normal counterparts. One example is a rare form of gout, in which the defective enzyme has an increased tendency to bind one of its chemical substrates.

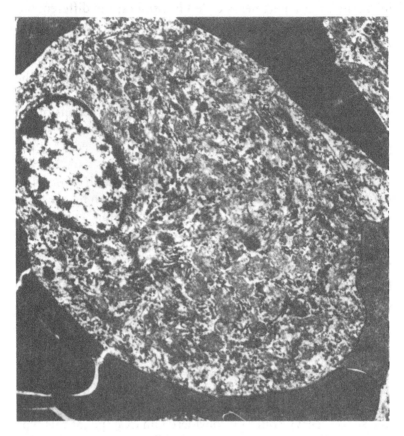

FIGURE 2.9 One of the most striking features of Gaucher disease is the presence of large, lipid-laden cells, called Gaucher cells, in the tissues of patients with the disorder. This photograph, taken through an electron microscope, shows a typical Gaucher cell magnified about 4,000 times. The cytoplasm has a "wrinkled tissue paper" or "crumpled silk" appearance. Gaucher cells are particularly numerous in the spleen, the liver, and the bone marrow. SOURCE: Reprinted, with permission, from R. O. Brady and F. M. King, "Gaucher's Disease," in H. G. Hers and F. Van Hoof, eds., *Lysosomes and Storage Diseases* (New York: Academic Press, 1973), p. 386.

Consequently, a normal biochemical reaction proceeds at a faster-than-normal rate, resulting in overproduction of uric acid. Accumulation of uric acid in the blood and in joints causes the clinical signs of gout.

Disruption of Cell and Organ Function

Every cell in the body (except red blood cells) carries a complete set of genetic instructions, but the manner in which these instructions are used is highly specialized. Muscle cells, for example, do not use the genes that code for hemoglobin; conversely, red blood cell precursors do not use the genes that code for the proteins of muscle fibrils. Also, cells in different tissues produce and utilize different classes of enzymes.

The impact of a single-gene disorder is determined in part by the function of affected cells and tissues. A defect that inactivates a liver enzyme has a very different clinical picture (for example, liver enlargement and hypoglycemia in the glycogen storage diseases) from a defect that inactivates a crucial enzyme in the brain (which may cause progressive motor weakness and developmental arrest, as in Tay-Sachs disease).

Tissue specificity also plays a major role in determining whether a disease will be amenable to gene therapy. Early gene therapy trials probably will be restricted to the insertion of healthy genes into bone marrow cells. Bone marrow cells give rise primarily to cellular components of the blood. Thus, genetic diseases that affect blood cell function will be among the most likely candidates for early gene therapy.

The feasibility of using gene transfer in bone marrow cells to correct disorders involving tissues other than blood depends on the answers to several questions:

- Is the gene product made by blood cells?
- Where does the gene product act, inside the cell or outside the cell?
- Is the affected organ accessible to gene products in the bloodstream?

The most suitable candidates for gene therapy among diseases not restricted to blood cells will be those involving "housekeeping enzymes." These are enzymes that are present in every cell in the body. The lack of a housekeeping enzyme may affect some cell types more than others (affected cells may be more susceptible to damage from a toxic substrate, or they may have a greater need for a missing end product). The reason for focusing on housekeeping enzymes is that scientists believe it may be easier to make a treated bone marrow cell produce a gene product if the product is one normally produced by healthy bone marrow cells and their descendants.

Successful gene therapy also requires access to the toxic substance responsible for a target disease. If insertion of a healthy gene enables bone marrow-derived cells from a patient to produce normal enzyme, but the substrate for that enzyme is locked inside liver or muscle cells, then the effort may not result in clinical improvement. Gene therapy will be effective in this situation only if the affected cells have an active transport mechanism that allows them to take up the normal enzyme (produced by blood cells carrying the inserted gene) from their environment.

If a toxic metabolite resulting from a genetic defect circulates in the bloodstream, treated bone marrow cells must be able to clear it from the circulation faster than it is produced by enzyme-deficient tissues. In this situation, the likelihood of clinical improvement with gene therapy depends in part on the level of enzyme expressed by the inserted gene. Animal studies indicate that cells carrying an inserted gene usually produce less of a desired enzyme than normal cells. Thus, gene therapy may be useful only for diseases in which low levels of circulating enzyme are corrective.

One of the key issues in gene therapy is whether the technique will be effective for genetic disorders that affect the central nervous system (CNS), such as Lesch-Nyhan syndrome (a devastating, X-linked disorder associated with severe gout-like symptoms, spastic cerebral palsy, mental retardation, and compulsive self-mutilation). The brain is separated from the circulation by a physiological barrier that prevents the passage of enzymes and other molecules. Clinical trials of enzyme replacement therapy underscore the difficulties imposed by this "blood–brain barrier" (see Chapter 4).

One hopeful sign in this regard is the possibility that some cells in the central nervous system, the microglial cells, may originate in the bone marrow.[3] In a study involving bone marrow transplantation in mice, researchers found microglial cells of donor origin in the central nervous system nine months after bone marrow transplantation. If microglial cells are of bone marrow origin, they could be a continuing source for the local production and release of normal enzymes in the brain.

Several clinical studies of bone marrow transplantation in patients with lysosomal storage disorders, described in Chapter 4, have produced encouraging results, but numerous additional studies will be required to determine the potential impact of gene therapy on the treatment of inherited disorders of the central nervous system.

Conclusions

Many different factors must be considered in the selection of diseases for early gene therapy trials. A potential candidate must be fatal early in life. It should have a fairly uniform clinical course that is not altered by more conventional therapeutic measures. In addition, researchers must know the identity of the abnormal gene and where it is expressed in the body.

Defects in "housekeeping genes" are the most likely candidates for early trials. Unlike genes that code for cell-specific proteins, such as hemoglobin, housekeeping genes may not require precise regulation. At present, scientists have very little control over the amount of protein produced by an inserted gene, and no control over when it is produced.

Current technologies require that gene insertion into cells take place outside the body, and that target cells be either bone marrow cells or skin cells. Early clinical trials probably will employ bone marrow cells. Thus, the most likely candidates for gene therapy are diseases that are clinically restricted to tissues derived from bone marrow. The efficacy of gene therapy for disorders affecting other organs and tissues will depend on whether enzymes produced by treated bone marrow cells or their descendants can gain access to toxic substrates.

Two rare inherited diseases of the immune system appear to satisfy many current requirements for gene therapy. They are adenosine deaminase (ADA) deficiency and purine nucleoside phosphorylase (PNP) deficiency.[4] Children with ADA deficiency have almost no specific immunity; those with PNP deficiency can produce antibodies but do not have cell-mediated immunity (required for the destruction of virus-infected host cells and other essential immune functions). Both disorders involve housekeeping enzymes, but their clinical presentations are restricted to bone marrow-derived white blood cells.

—— ACKNOWLEDGMENTS ——

Chapter 2 is based on the presentations of Rochelle Hirschhorn, David W. Martin, Jr., Robertson Parkman, Leon E. Rosenberg, and James B. Wyngaarden.

—— NOTES ——

1. The prospects for gene therapy for beta thalassemia improved recently, when scientists demonstrated in animal studies that an intact beta-globin gene could be introduced into bone marrow stem cells and expressed at reasonable levels in the appropriate cell types (see Chapter 6).

2. As explained in Chapter 4, the status of ADA deficiency as a potential candidate for early human studies of gene therapy depends in part on the outcome of clinical trials with a new form of enzyme replacement therapy. If the enzyme replacement therapy is successful, the relative benefits and risks of gene therapy for ADA-deficient patients will change.

3. The origin of microglial cells remains highly controversial. Some researchers believe they are derived from hematopoietic tissue, while others believe they are of neuroectodermal origin.

4. The ultimate choice of disease candidates for early clinical trials of gene therapy will depend on existing knowledge in many different areas of medical science. See, for example, the discussion of enzyme replacement therapy in ADA deficiency (Chapter 4).

———— SUGGESTED READINGS ————

Anderson, W. French. 1985. "Beating Nature's Odds." *Science 85,* November, pp. 49–50.

Drlica, Karl. 1984. *Understanding DNA and Gene Cloning: A Guide for the Curious.* New York: John Wiley.

The New Human Genetics: How Gene Splicing Helps Researchers Fight Inherited Disease. 1984. Written by Maya Pines. Bethesda, MD: National Institute of General Medical Sciences, National Institutes of Health. NIH Publication no. 84-662.

Parkman, Robertson. 1986. "The Application of Bone Marrow Transplantation to the Treatment of Genetic Diseases." *Science* 232:1373–1378.

Schmeck, Harold M., Jr. 1985. "The Promises of Gene Therapy." *New York Times Magazine,* November 10, pp. 116–119.

Stanbury, John B., James B. Wyngaarden, Donald S. Fredrickson, Joseph L. Goldstein, and Michael S. Brown. 1983. "Introduction." *The Metabolic Basis of Inherited Disease,* pp. 3–38. 5th edition. New York: McGraw-Hill.

You Have a Right to Know about Jewish Genetic Diseases. Pamphlet produced by the National Foundation for Jewish Genetic Diseases, Inc., 250 Park Avenue, Suite 1000, New York, N.Y.

3 | Diagnosis of Genetic Disorders

Gene therapy represents an exciting new direction in medicine, but its ultimate impact will depend on how it compares with other methods of coping with genetic diseases. This chapter focuses on genetic diagnosis, a broad field that encompasses prenatal diagnosis, newborn screening, and techniques to identify healthy carriers of genetic disorders, as well as the diagnosis of inherited diseases in sick children and adults. Consider the following examples:

- A young couple whose first baby succumbs to liver failure resulting from the genetic disease alpha$_1$-antitrypsin deficiency desperately want another child, but they are equally determined to avoid the pain of watching a second child die. At first, they decide not to have any more children. Then they learn about a new test that can identify alpha$_1$-antitrypsin deficiency in the fetus. The woman becomes pregnant and, at the appropriate time, travels to a neighboring city to undergo the test. Within days, physicians tell the expectant parents that the fetus does not have the hereditary disease. Six months later, the woman gives birth to a healthy baby girl.
- A 20-year-old woman tells her fiance that she does not want to have children. Her younger brother has Duchenne muscular dystrophy (DMD), and she is worried that she may pass the disease on to her own sons. Her fiance reads about a diagnostic test that can identify

carriers of the DMD gene. A subsequent family study shows that the woman does not carry the disease.

- A pediatrician receives a call from a laboratory in the Department of Public Health informing her that one of her patients has tested positive on a newborn screening test for congenital hypothyroidism. Further tests reveal that the infant's condition is caused by a genetic inability to produce thyroid hormone. The physician prescribes regular treatment with the hormone, and the baby grows and develops normally. Without treatment, the child would have developed the classic signs of cretinism, including severe mental retardation.

- A young executive whose father died of a heart attack in his late forties worries about his own risk of an early death. He contacts a company that uses a battery of new diagnostic tests to assess his genetic make-up. The tests indicate that the young man has inherited the tendency for heart disease and that he also has a higher-than-average risk of developing diabetes. The company provides him with a nutritional plan and other information designed to reduce his risk of disease.

Genetic diagnosis has done more to reduce the human suffering associated with inherited diseases than any other medical technology. Prenatal diagnosis and carrier screening programs have provided information for thousands of couples facing difficult choices about conceiving a child. Through newborn screening programs, physicians have identified numerous children with treatable genetic diseases in time to prevent mental retardation and other permanent disabilities.

Recent advances in recombinant DNA technology, which allow researchers to look directly at altered DNA structure, have greatly improved the range and reliability of diagnostic tools. Physicians can now detect more than 200 inherited diseases in the womb. But the new technology also has had another effect—it has broadened the focus of genetic diagnosis. Increasingly, researchers are turning their attention to the genetic basis of conditions not usually thought of as inherited diseases. For example, scientists have identified genetic markers that signal increased susceptibity to heart disease and certain types of cancer. The final section of this chapter briefly examines the potential uses and abuses of these new predictive tests.

Childbearing Decisions

In June 1980, Bernadette Modell and her coworkers at the University College Hospital Medical School in London described the effect of introducing prenatal diagnosis on the reproductive behavior of couples at risk of having a child with severe beta thalassemia. Prior to the availability of prenatal diagnosis, couples who knew they were at risk (either because they already had a child with thalassemia or because both parents tested positive on carrier screening tests) virtually stopped having children. Most pregnancies that occurred were accidental, and 70 percent of couples chose to abort the fetus rather than run the risk of producing another affected child.

Prenatal diagnosis changed the situation dramatically: reproductive patterns in the study group returned almost to normal. Fewer than 30 percent of subsequent pregnancies were terminated because of thalassemia.

Similar results have been reported for many other disorders. Following the introduction of prenatal diagnosis and carrier screening tests, large numbers of normal pregnancies have been started and continued by high-risk couples who previously had decided to forgo having children.

Prenatal Diagnosis

Widespread use of prenatal diagnosis began in the early 1970s, following the discovery that cells floating in the amniotic fluid surrounding a fetus could be used to detect major chromosomal abnormalities and selected biochemical defects. Today, physicians employ several different procedures for collecting fetal cells, a variety of analytic techniques, precise methods for assessing fetal structural abnormalities, and a maternal blood test that classifies risk for major central nervous system disorders.

Most of these diagnostic tools are used selectively: physicians recommend them only if maternal age, family history, or population studies (for disorders that occur with greater frequency in certain ethnic groups) suggest that a particular fetus is at risk. Also, the tests are very specific. A negative result on a prenatal test for galactosemia, for example, does not guarantee that a fetus is healthy—it indicates only that the fetus does not have galactosemia.

Amniocentesis. The most common method of obtaining fetal cells for analysis is amniocentesis, in which the physician inserts a needle through the mother's abdominal wall to withdraw a small amount of amniotic fluid. The usual procedure is to locate the placenta and determine the position of the fetus with ultrasound (a technique using high-frequency sound waves to produce body images). This prevents injury to the fetus and placenta when the needle enters the amniotic sac.

The majority of those who seek amniocentesis are women concerned about Down syndrome. For women between the ages of 25 and 30, the risk of having a child with Down syndrome is 1 in 1,000; at age 35 the risk increases to 1 in 350, and by age 40 the risk is 1 in 100. Dr. Maurice Mahoney, a professor of human genetics, pediatrics, and obstetrics and gynecology at Yale University School of Medicine, estimates that 65 percent of pregnant women in New England who are 35 years of age or older elect to undergo amniocentesis.

Early efforts to use amniocentesis to diagnose genetic diseases other than chromosomal disorders were limited by the nature of the cells that were collected. The cells come primarily from the fetal membranes, the fetal skin, and, to a lesser extent, the gastrointestinal, respiratory, and urinary tracts. Researchers could diagnose inborn errors of metabolism only if the errors altered biochemical processes normally occurring in the cell types available. Thus, they could diagnose Tay-Sachs disease (because amniotic cells from normal fetuses produce the enzyme hexosaminidase A and amniotic cells from affected fetuses do not), but they could not diagnose the common blood disorder thalassemia (because amniotic cells do not produce hemoglobin proteins).

Fetoscopy and Fetal Blood Sampling. The development of fetoscopy, a technique to visualize the fetus directly using an instrument called a fiberoptic endoscope, significantly broadened the scope of prenatal diagnosis. In 1974, researchers adapted fetoscopy to permit sampling of fetal blood. For the first time, physicians could diagnose thalassemia and sickle cell disease in the fetus. Subsequently, laboratory procedures were developed to detect other genetic disorders affecting blood and blood products, including hemophilia and chronic granulomatous disease (an X-linked disorder in which affected males are highly susceptible to bacterial infections).

More recently, physicians have developed the capability to sample fetal blood without fetoscopy. Using ultrasound guidance, they insert a needle through the mother's abdomen directly into the large blood vessels in the umbilical cord.

The principal drawback of fetal blood sampling is its relatively high complication rate. The risk of miscarriage after fetal blood sampling is about 2 percent; the risk of fetal loss after amniocentesis is only 0.5 percent (or about 1 in 200 women tested).

Chorionic Villus Sampling. One problem shared by amniocentesis and fetal blood sampling is that they cannot be performed until after the fourteenth week of pregnancy (fetal blood sampling usually takes place between weeks 18 and 21). This has been a particular problem for amniocentesis, because amniotic cells must be grown in the laboratory for one to three weeks before scientists have enough cells for most conventional analytic procedures. Thus, test results usually are not available until well into the second trimester. Abortion of an affected fetus at that time may be very difficult emotionally for the parents, and is much harder physically for the mother than an abortion during the first trimester.

To overcome this difficulty, medical geneticists have developed an alternative to amniocentesis called chorionic villus sampling, or CVS. Chorionic villi are fingerlike projections of the membrane surrounding the embryo early in pregnancy. They participate in the exchange between maternal and fetal circulation and gradually evolve into the mature placenta.

Chorionic villus cells carry the same genetic information as the developing fetus.

CVS is performed by either of two techniques during the ninth through eleventh weeks of pregnancy. The first technique involves the insertion of a catheter through the vagina and cervix into the uterus. Using ultrasound guidance, the physician places the tip of the catheter into the substance of the chorion and then applies suction to tear off a small sample of villi. The sample is cleaned of adhering maternal cells and then either analyzed directly or prepared for cell culture. The second technique more closely resembles amniocentesis—chorionic villi are collected via a needle inserted through the mother's abdominal wall.

CVS yields more live cells than amniocentesis, so technicians often can conduct necessary tests without waiting for cells to grow in the laboratory. The combination of an earlier procedure and faster results may save up to two months of waiting time for anxious parents.

Despite these benefits, CVS still is not universally available in the United States, in part because the risk of miscarriage following CVS, about 2 percent, is significantly higher than that associated with amniocentesis. Researchers believe that this risk will drop as physicians become more experienced with the procedure. For now, many physicians recommend that its use be limited to women who already have given birth to a child with a serious genetic disease.

New Analytic Techniques. The most important advance in the field of prenatal diagnosis over the past decade has been the use of recombinant DNA technology to analyze fetal cells. Conventional analytic tests focus on changes at the protein level: researchers measure enzyme activity in amniotic cells or search for abnormal protein molecules in the blood. These tests are useful only if scientists know the biochemical basis of the disease in question. Also, affected cells must be accessible by one of the standard cell collection procedures.

In contrast, new tests using recombinant DNA technology focus directly on the structure of the human genome (see Table 3.1). For example, conventional diagnostic procedures are not

TABLE 3.1. Diagnosis of genetic disorders by recombinant DNA methods.

Disorder	Method	Clinical use[a]
Adenosine deaminase deficiency with severe combined immune deficiency	D[b]	Limited availability
Adult polycystic kidney disease	RFLP[c]	Limited availability
Alpha and beta thalassemia	D and RFLP	Relatively routine
Alpha$_1$-antitrypsin deficiency	D	Limited availability
Chronic granulo-matous disease	D and RFLP	Limited availability
Cystic fibrosis	RFLP	Relatively routine
Duchenne muscular dystrophy	RFLP	Relatively routine
Familial hyper-cholesterolemia	D and RFLP	Limited availability
Hemophilia A	D and RFLP	Relatively routine
Hemophilia B	RFLP	Relatively routine
Huntington disease	RFLP	Use limited to research protocols
Lesch-Nyhan syndrome	D and RFLP	Relatively routine
Ornithine trans-carbamylase deficiency	RFLP	Limited availability
Phenylketonuria	D	Limited availability
Retinoblastoma	D and RFLP	Relatively routine
Sickle cell disease	D	Relatively routine

a. Clinical practice is changing very rapidly in this field. Tests that are now available only on a limited basis may become relatively routine within a year.

b. D = direct assay for relevant gene.

c. RFLP = test based on identification of restriction fragment length polymorphisms in family (see text).

SOURCE: Katherine Wood Klinger, Ph.D., Manager of Genetic Disease Research, Integrated Genetics, Framingham, Massachusetts.

effective in the diagnosis of phenylketonuria (PKU), because the abnormal enzyme that causes the disease is produced almost exclusively by liver cells (which cannot be collected by amniocentesis, fetal blood sampling, or CVS). However, all cells in an affected fetus carry the abnormal gene that codes for the enzyme. Recombinant DNA techniques allow physicians to determine which fetuses in affected families carry the normal gene and which carry the abnormal gene.

The most important tools in recombinant DNA technology are bacterial proteins called **restriction enzymes.** These enzymes cut DNA into fragments. Many other substances also break up DNA—the feature that makes restriction enzymes unique is their specificity. Each restriction enzyme recognizes and cuts a particular sequence in the DNA molecule. For example, the restriction enzyme *Eco*RI recognizes the DNA sequence GAATTC (in which G represents the nucleotide base guanine, A represents adenine, T represents thymine, and C represents cytosine) and cuts it between the G and the A; the restriction enzyme *Mst*II recognizes the sequence CCTNAGG (the N can be any one of the four nucleotides) and cuts it between the second C and the T.

When a restriction enzyme is used to cut a piece of DNA, the lengths of the resulting fragments depend on the distances between recognition sites (the sites at which it cuts). Scientists determine the size of each fragment with a process called electrophoresis. (An electric field is used to make DNA fragments move through a special gel. Smaller fragments move through the gel faster than larger fragments, so researchers can determine the length of any fragment by comparing its position on the gel to that of a DNA molecule of known size.)

The use of restriction enzymes in prenatal diagnosis is based on the discovery of inherited variations in restriction enzyme recognition sites in human DNA. An enzyme that cuts a piece of DNA from one person into three fragments, for example, may cut a corresponding piece of DNA from another person into only two fragments. The reason for the variation (called a restriction fragment length polymorphism, or RFLP) is that the second person is missing a recognition site, perhaps because of a change in a single nucleotide (see Figure 3.1).

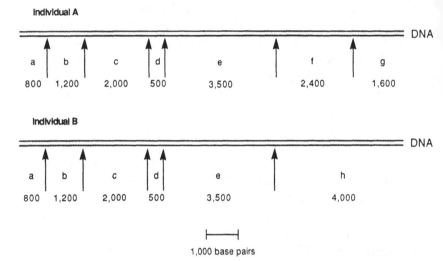

FIGURE 3.1 When a specific restriction enzyme cuts DNA, it may produce fragments of different sizes in the DNA of different people. For example, *Eco*RI will cut DNA wherever the base sequence GAATTC occurs (represented here by vertical arrows). Individual **A** has six of these recognition sites, yielding fragments **a** through **g**, but individual **B** has only five recognition sites. Thus, the length of fragment **h**—measured in DNA base pairs—from individual **B**'s DNA equals the sum of the lengths of fragments **f** and **g** produced by *Eco*RI from individual **A**'s DNA. Such natural variations, or polymorphisms, in restriction fragment lengths are inherited. SOURCE: Adapted from *The New Human Genetics: How Gene Splicing Helps Researchers Fight Inherited Disease*, written by Maya Pines (Bethesda, MD: National Institute of General Medical Sciences, 1984), NIH Publication no. 84-662.

The variations in DNA that produce restriction fragment length polymorphisms rarely cause disease, but in some cases they can be used as indirect markers for specific genetic disorders. If a restriction fragment length polymorphism is very near a disease-producing gene on a chromosome, the two have a high probability of being inherited together.

The association of an RFLP with a disease-producing gene is determined by a technique called **linkage analysis.** Researchers use one or more restriction enzymes to cut DNA from both healthy and affected members of a family with a given genetic disease. The goal is to find a DNA pattern that is linked to the disease. If such a pattern is found, it can be used to distinguish healthy fetuses from those who have inherited the defective gene even if the disturbance at the protein level is not known.

A new linkage analysis test for Duchenne muscular dystrophy (DMD) recently led to the births of 16 healthy baby boys who probably would not have been born if the test had not been available. The mothers of these infants had been identified as carriers of the debilitating and ultimately fatal X-linked muscle disorder by researchers at Baylor College of Medicine in Houston, Texas. In the past, many women with family histories of DMD chose to abort all male fetuses rather than run the risk of bearing an affected son. The new prenatal test indicated that the boys did not have the genetic marker for DMD and gave their parents the confidence to continue the pregnancies.

The accuracy of these diagnostic tests depends on how close the genetic marker (the locus of the RFLP) is to the defective gene. The DMD prenatal test is estimated to have an accuracy level of between 90 and 99 percent.

If a genetic disease is caused by a mutation that actually alters a restriction enzyme recognition site, family studies may not be necessary. For example, the mutation responsible for sickle cell disease on the beta-globin gene happens to occur right in the middle of a recognition site for the restriction enzyme *Mst*II. In laboratory tests, this enzyme cuts the normal beta-globin gene into two fragments, but leaves the sickle cell gene intact (see Figure 3.2). This allows researchers to distinguish between the two quite easily.

In a 1983 review of prenatal diagnosis, Charles Epstein and his coworkers from the University of California at San Francisco noted, "In the short time that this [*Mst*II] assay has been available, it has proven to be a rapid, sensitive, and accurate technique for the prenatal diagnosis of sickle cell anemia. Sufficient fetal DNA can be extracted directly from the fetal cells present in 8 ml [about one quarter of an ounce] of amniotic fluid, thereby eliminating the need to grow the

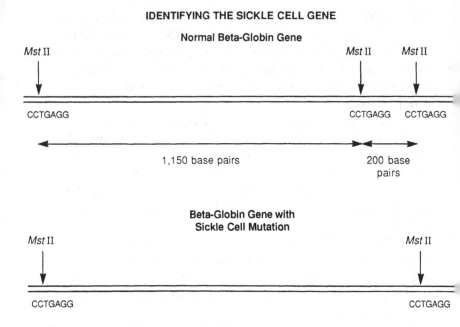

FIGURE 3.2 Several different methods are available for the prenatal diagnosis of sickle cell disease: one involves the restriction enzyme *Mst*II. When the beta-globin gene is normal, *Mst*II cuts through it at the sequence CCTGAGG, producing two fragments, 1,150 and 200 base pairs in length. However, the sickle cell mutation converts the DNA sequence at this point to CCTGTGG, thereby abolishing the *Mst*II recognition site; this results in just one fragment, 1,350 base pairs in length. SOURCE: Adapted from *The New Human Genetics: How Gene Splicing Helps Researchers Fight Inherited Disease*, written by Maya Pines (Bethesda, MD: National Institute of General Medical Sciences, 1984), NIH Publication no. 84-662.

cells in culture. This method not only reduces the cost of the test, but makes it possible to obtain a result routinely in as short a time as two weeks."[1]

Oligonucleotide Analysis. An alternative technique for diagnosing disorders caused by point mutations employs pieces of synthetic DNA called oligonucleotides. These short nucleotide chains, constructed out of laboratory chemicals, bind to matching human DNA. They are used as "probes" to distinguish between normal and abnormal genes.

A new diagnostic test for a form of beta thalassemia common among people of Mediterranean origin demonstrates how this technique is used. Scientists employ two different radiolabeled oligonucleotides—one exactly matches the abnormal nucleotide sequence in the defective beta-globin gene, and the other matches its normal counterpart. The probes are added individually to DNA from fetal cells (collected by CVS or amniocentesis). If the probe corresponding to the abnormal gene segment sticks to fetal DNA and the other probe does not, the fetus will develop thalassemia; if both probes stick, the fetus has inherited one normal and one abnormal gene and will be a carrier like its parents. A similar test has been devised for the disease alpha$_1$-antitrypsin deficiency, which predisposes affected individuals to severe and often fatal liver disease in infancy and leads to obstructive emphysema in early adult life.

The ability to diagnose hemoglobin disorders and alpha$_1$-antitrypsin deficiency in cells obtained by CVS or amniocentesis represents a significant advance over conventional methods.

Ultrasound. Advances in ultrasound technology over the past decade have kept pace with the dramatic changes in other areas of prenatal diagnosis. High-resolution, real-time images have replaced the fuzzy "snapshots" produced by ultrasound equipment of the 1970s.

Physicians who have a firm knowledge of normal fetal anatomy and development can use the high-resolution images to detect a variety of skeletal disorders, including limb malformations and osteogenesis imperfecta (a generalized disorder of connective tissue involving bone, but also affecting tendons, ligaments, teeth, and other structures). Ultrasound imaging of the central nervous system reveals defects such as anencephaly (an extremely small head size associated with absence of part of the brain) and hydrocephalus (abnormal accumulation of fluid in the brain, see Figure 3.3). Kidney abnormalities and urinary tract obstructions also are readily detectable.

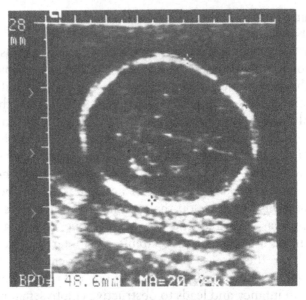

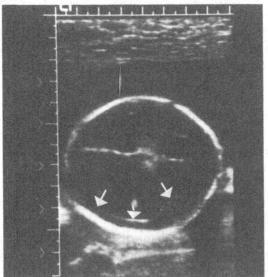

FIGURE 3.3 Ultrasound images showing the head of (a) a normal 20-week-old fetus, and (b) a 22-week-old fetus with severe hydrocephalus (distension of the ventricles, or spaces, in the brain resulting from an abnormal accumulation of cerebrospinal fluid, CSF). The arrows in (b) indicate the border between the walls of the lateral ventricle and surrounding brain tissue. The brain tissue has been compressed and distorted by the excess pressure. SOURCE: Dr. Frederick Doherty, Director, Ultrasound Division, New England Medical Center, Boston.

When fetal ultrasound is combined with standard echo-cardiography techniques, physicians can assess the structure and function of the fetal heart. To evaluate fetal liver function, researchers use ultrasonography to guide a biopsy needle into the fetal liver.

Maternal Serum Alpha-Fetoprotein Screening. All of the procedures described so far focus on the fetus or fetal tissues. Another addition to prenatal screening involves a maternal blood test.

In June 1983, the U.S. Food and Drug Administration authorized the distribution to medical laboratories of test kits for measuring alpha-fetoprotein (AFP) levels in the blood of pregnant women. AFP is a protein found in the serum of all pregnant women. Repeated studies have shown, however, that women carrying fetuses with spina bifida (in which the spinal column fails to close during development) and related defects of the central nervous system have higher levels of AFP in their blood than women carrying healthy fetuses.[2] In addition, reports from large screening programs in England, Germany, and the United States indicate that pregnancies in which the fetus has Down syndrome are associated with lower-than-expected AFP levels.

In many parts of the United States, the maternal serum alpha-fetoprotein (MSAFP) screening test has become a routine part of prenatal care. Figure 3.4 shows the expected outcome of screening for a random group of 1,000 pregnant women under age 35 (most women over age 35 are offered amniocentesis irrespective of MSAFP screening), of whom roughly 100 will have MSAFP levels that call for further investigation. Among the 50 women who have higher-than-normal MSAFP levels, further tests will show that two have fetuses with structural defects involving the central nervous system or some other organ system in the body. Another 28 fetuses will appear normal (based on ultrasound evaluations and other diagnostic studies), but may be at increased risk of future problems. Dr. Maurice J. Mahoney of Yale University School of Medicine explains that the risk of a poor pregnancy outcome (premature birth, growth retardation in the womb,

MATERNAL SERUM ALPHA-FETOPROTEIN SCREENING

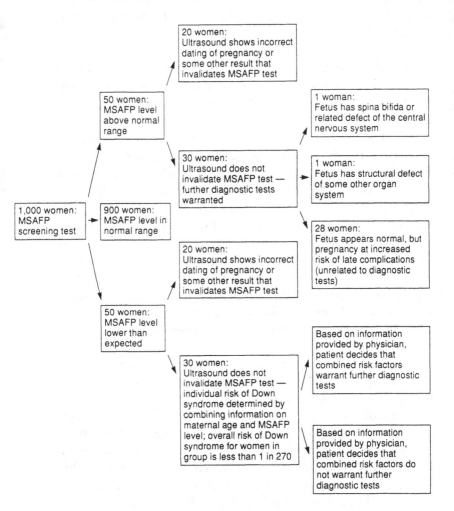

FIGURE 3.4 Flow chart showing the expected results of screening a random sample of 1,000 pregnant women under age 35 with the maternal serum alpha-fetoprotein (MSAFP) test. (Most women age 35 and older are offered amniocentesis irrespective of MSAFP screening.) SOURCE: Dr. Maurice J. Mahoney, Yale University School of Medicine.

or stillbirth) is two to three times higher for women with elevated MSAFP levels than for women with normal MSAFP levels. Such women require careful monitoring—in some cases, physicians may be able to intervene to prevent serious problems later in pregnancy.

A lower-than-expected MSAFP test result indicates an increased risk of Down syndrome, but the size of the risk depends on the mother's age. The odds of having a child with Down syndrome for a 20-year-old woman are about 1 in 1,240. A low MSAFP level increases the odds to about 1 in 600—or about the same risk as an average woman in her early thirties. In contrast, a low MSAFP level in a woman who is 34 years old increases the odds that she will have a child with Down syndrome from about 1 in 350 to greater than 1 in 150.

The MSAFP test is an important screening tool, but its widespread availability raises many practical and ethical questions, especially about the risk of unnecessary abortions. False positive results may be caused by incorrect dating of the pregnancy (one week can make a difference), variations in maternal weight gain, maternal diabetes, and other medical conditions. Thus, women who test positive need appropriate follow-up diagnostic procedures (including ultrasound to reassess gestational age and, if necessary, amniocentesis), combined with knowledgeable genetic counseling.

At the moment, maternal serum AFP screening is offered by a patchwork of hospital, academic, and commercial laboratories. Not all of these facilities have the capability to participate in the coordination of follow-up care. More importantly, some may lack the skills necessary to help physicians accurately interpret test results for their patients.

Public health studies indicate that the concept of risk is very difficult to convey to a lay audience. For example, if a mother is told that the results of her AFP screening test indicate that her risk of having a Down syndrome pregnancy is greater than 1 in 270 (the current cut-off for recommending further tests), she may become very anxious. Despite the still low probability that her fetus is affected, it is conceivable that she might decide to forgo an amniocentesis and have an abortion.

This situation provides one of the best illustrations of the challenges created by recent advances in health care technology. Physicians must convey complex information in an unbiased fashion to patients who already face many new and unfamiliar choices.

Carrier Screening

In the past, most couples who sought prenatal diagnosis for an inherited metabolic disorder did so because a previous pregnancy had resulted in the birth of an affected child. With the advent of screening tests for heterozygous carriers of autosomal recessive and X-linked disorders, some high-risk couples can be detected and counseled about reproductive choices before a first pregnancy.

The most successful example of prospective carrier screening has been the effort to reduce the incidence of Tay-Sachs disease. In 1970, between 50 and 100 babies were born with Tay-Sachs disease in the United States; today, the total has dropped to 10 or below.

The development of a preventive program for Tay-Sachs disease was possible for three reasons. First, it occurs mainly in a defined population—the frequency of carriers is 10 times higher among Jews of European ancestry than it is in the general population. Second, scientists have a simple and inexpensive test to identify carriers; and third, prenatal diagnosis gives at-risk couples (in which both partners are carriers) the option of having only normal children. To date, Tay-Sachs screening programs have detected more than 15,000 carriers (about 1 in every 25 Jewish people tested) and identified more than 800 couples at risk.

Screening programs also have been developed for sickle cell anemia and for certain forms of thalassemia. In general, however, tests for carriers of autosomal recessive and X-linked disorders are employed on an individual basis as part of a comprehensive genetic counseling program. For example, each sister of a boy born with Lesch-Nyhan syndrome has a 50 percent chance of inheriting the defective gene and passing it on to one or more of her own sons. Tissue culture studies of the skin cells of such a woman can determine whether or not she is a carrier.

Conventional tests for heterozygous carriers, like conventional tests used in prenatal diagnosis, focus on changes at the protein level. New tests using recombinant DNA technology focus directly on DNA structure. Scientists employ restriction enzymes and DNA probes to search for DNA patterns that distinguish reliably between family members who have a genetic disease and those who do not. Once such a pattern has been established, it can be used to identify carriers even if the mutant protein remains a mystery.

In the Duchenne muscular dystrophy example described earlier, the women identified as DMD carriers came from a group of 123 women who had at least one relative with Duchenne muscular dystrophy.[3] In the past, all 123 might have undergone prenatal diagnosis and considered the termination of a male pregnancy. With the new screening technique, 41 women were found to be at very low risk of being a carrier. A similar technique recently became available to screen suspected carriers of cystic fibrosis, the most common genetic disease among Caucasians in the United States.

Fetal Therapy

Although prenatal diagnosis and carrier screening are used primarily to provide couples with information on which to base reproductive choices, a new use is developing. In a small number of cases, physicians can use the diagnostic information acquired through ultrasound or amniocentesis to develop plans for medical intervention inside the womb.

The field of fetal therapy is still in its infancy. Some of the procedures used are direct adaptations of diagnostic techniques. For example, the technique used to sample fetal blood (insertion of a needle into blood vessels in the umbilical cord) has been adapted to administer red blood cells to fetuses with severe anemia.

The widespread use of fetal heart rate monitoring and sophisticated ultrasound techniques has greatly improved the management of cardiac problems in the fetus, especially problems associated with irregular heart rhythms. While most disturbances of fetal heart rhythm are of little clinical importance and pose no danger to the fetus, some may be an indication of impending fetal heart failure. In such cases, physicians often can save an endangered fetus by administering antiarrhythmic drugs to the mother.[4]

Increased familiarity with the intrauterine environment also has led to several forms of fetal surgery. In July 1986, representatives of the International Fetal Surgery Registry reviewed 114 reports of fetal surgery from 21 medical centers.[5] Seventy-three cases involved attempts to overcome life-threatening obstructions of the fetal urinary tract (obstructive hydronephrosis) by implanting catheters to drain urine into the amniotic fluid. The remaining 41 procedures were performed to drain fluid from the dilated ventricles (spaces in the brain) of fetuses with hydrocephalus. Procedure-related fetal death rates were 4.2 percent and 9.8 percent, respectively.

The registry data suggest a "possible benefit of prenatal therapy in selected cases of fetal obstructive hydronephrosis," but are not encouraging about the use of fetal surgery to relieve hydrocephalus. The authors emphasize the need for controlled clinical trials to fully assess the efficacy and safety of these techniques.

Fetal therapy for inherited metabolic disorders has progressed more slowly than efforts to correct physical defects. Two exceptions are vitamin B_{12}-responsive methylmalonic acidemia and biotin-responsive carboxylase deficiency. Both of these disorders result from defects in vitamin metabolism, and both have been treated by the administration of pharmacologic doses of the vitamins to mothers carrying affected fetuses.

The future of fetal therapy depends, in part, on the development of successful techniques for correcting genetic defects after birth. Theoretically, enzyme replacement therapy, bone marrow transplantation, and gene therapy could be employed to help a developing fetus if researchers knew that a particular defect caused irreversible damage in the womb. In practice, however, these procedures will have to be shown to be safe and effective in infants and children before being adapted for use in the fetus.

Newborn Screening

Researchers have estimated that specific treatment techniques provide complete relief for about 12 percent of hereditary metabolic diseases. The goal of newborn screening, which is the most widespread form of genetic diagnosis in the United States, is to identify children who could benefit from such measures before permanent damage occurs.

Newborn screening began in the early 1960s, following the development by Dr. Robert Guthrie of an inexpensive and simple blood test for the disease phenylketonuria (PKU), a cause of severe mental retardation. Ten years earlier, Dr. Horst Bickel and his associates had described a dietary treatment for PKU, but at that time few affected children were identified in time to prevent irreversible brain damage.

Today, PKU screening is mandatory in most industrialized nations and in some parts of the developing world. More than 100 million newborns have been tested, and about 10,000 affected children have been placed on the restricted diet. Numerous studies have shown that the cost of screening is more than outweighed by the savings in institutionalization and health care costs for severely disabled children.

The PKU test is performed on a small sample of blood, which is collected on filter paper and then mailed to a laboratory. In the late 1960s and early 1970s, Dr. Guthrie demonstrated that the same sample of dried blood could be used to screen for more than 10 other disorders, including maple syrup urine disease (an inherited disorder of amino acid metabolism named for the characteristic odor of the urine produced by affected babies), galactosemia (an inability to metabolize galactose), and homocystinuria. All three of these metabolic defects cause mental retardation and other serious complications, and all can be completely or partially controlled by diet.

States and regions vary somewhat in the diseases included

TABLE 3.2. Current newborn testing program (1987).

Specimen	Age obtained	Disorder screened	State Mass.[a]	State New York[b]	Test used
Newborn dried blood	3–5 days	Phenylketonuria (PKU)	Yes	Yes	Bacterial inhibition assay
		Maple syrup urine disease	Yes	Yes	Bacterial inhibition assay
		Homocystinuria	Yes	Yes	Bacterial inhibition assay
		Galactosemia	Yes	Yes	Enzyme assay
		Congenital hypothyroidism[c]	Yes	Yes	Radioimmunoassay
		Biotinidase deficiency	Yes	Yes	Enzyme assay
		Sickle cell disease	No	Yes	Electrophoresis
Dried urine	3–4 weeks	Amino acid disorders and organic acid disorders	Yes[d]	No	Efron method with paper chromatography

a. SOURCE: Newborn Screening Program, State Laboratory Institute, Jamaica Plain, MA 02130.
b. SOURCE: Newborn Screening Program, Wadsworth Center for Laboratories and Research, State of New York Department of Health, Albany, New York 12201.
c. About 20 percent of cases are familial in origin.
d. Available from State Laboratory Institute, but not mandated by law.

in their mandatory screening programs. Table 3.2 outlines the newborn testing programs in Massachusetts and New York. The blood test for hypothyroidism allows prompt administration of thyroid hormone to infants who otherwise would develop the classic signs of cretinism, including mental retardation and deafness. Urine screening is performed on samples obtained by placing a piece of filter paper in the diaper three to four weeks after birth. Parents send dried urine specimens to the laboratory and are notified by their physicians of any abnormal results.

Rapid advances in technology will soon make it possible to screen for cystic fibrosis, Duchenne muscular dystrophy, congenital adrenal hyperplasia (in which cortisol deficiency leads to life-threatening medical crises in early infancy, as well as abnormal development of the genitalia in females and early puberty in males), and many less common diseases. Public health officials, legislators, and the general public must weigh humanistic and economic considerations in deciding which tests are desirable for mass screening and how they should be integrated into current public health programs.

Detection of Autosomal Dominant Disorders

Perhaps the most publicized new test in medical genetics is the recombinant DNA technique that identifies persons who will develop Huntington disease, an incurable neurological disorder characterized by uncontrollable physical movements, loss of speech, and progressive mental deterioration. The average age of onset is 38 years. Long before death, which usually occurs within 10 to 15 years, patients become bedridden and totally unaware of their surroundings.

One of the many tragedies of Huntington disease is the terrible uncertainty that plagues affected families. Because it is an autosomal dominant disorder, each offspring of a person who has the disease has a 50 percent chance of inheriting the defective gene. Until recently, no one had a way of predicting which children would be affected and which would be spared.

In 1983, Dr. James Gusella of the Massachusetts General Hospital in Boston discovered a genetic marker for Huntington disease on human chromosome 4. Although scientists still do not know the identity of the Huntington disease gene, they can now identify affected individuals in families that have enough living members to conduct linkage analysis studies. Prenatal diagnosis is available for affected couples who want to have children but are unwilling to risk passing the disease on to the next generation.

Between 60 and 80 percent of the 100,000 Americans at risk of Huntington disease say they would like to take advantage of the new test to plan for the future.[6] Others believe, however, that uncertainty is better than the despair that would come from knowing that they carry the defective gene. Medical centers that offer the test are acutely aware of the need for extensive counseling, both before and after testing. Suicide is a major concern—the suicide rate among Huntington disease patients after symptoms appear is 8 percent.

In one sense, the situation with Huntington disease provides an important model for the future of medical genetics. An article in the *New York Times* of August 19, 1986, notes that at least 20 biotechnology companies are developing or planning to develop new genetic tests. Many of these companies plan to leave the realm of inherited single-gene defects and to focus instead on DNA markers linked to common disorders such as diabetes, heart disease, and major forms of mental illness. Some of the new tests will alert people to lifestyle changes that could reduce their risk of disease, but others will offer only advance warning of a potential disability or early death. Researchers need to learn more about the psychological effects of such warnings and how they influence major life decisions.

Predicting Lifetime Health Prospects

Public health specialists and others worry that society has not spent enough time considering how the new predictive genetic screening tests will be used. The tests could be used solely on an individual basis to help people alter their lifestyles in a manner consistent with long-term health goals. Or, they could be used to detect conditions that might be aggravated by substances in the workplace.

The major concern is that if the results of genetic screening tests are accessible to employers, insurance companies, and other large institutions in society, they could become a major factor in decisions about education, employment, insurance coverage, and related matters. For example, a company might not want to invest large amounts of time and money training someone for a high-level job if the person is likely to have a heart attack before age 45; conversely, it might favor such employees in jobs that require little training, because of the money that could be saved in pension benefits.

Scientists fear that the concepts of risk and uncertainty may be lost in the rush to apply the new genetic tests to modern life. Most serious diseases in adults result from complex interactions between several different genes and the environment. Thus, the presence of one abnormal gene might have only a small impact on the likelihood that a person will develop a particular disease. Who will determine the boundaries between an acceptable risk and an unacceptable risk?

Another major concern is that the new genetic tests will be adopted before they have been thoroughly evaluated. Proving the validity of the new predictive tests for heart disease, diabetes, cancer, and similar disorders could take decades.

Widespread public understanding of the potential advantages and disadvantages of the new predictive tests is essential. Well-designed genetic screening programs, employing accurate diagnostic tests and knowledgeable counseling, could greatly expand opportunities for people to make informed decisions about their own health and medical care. The same tests, used for mass screening of potential job applicants or other large groups, could impinge on basic human rights, including the right to privacy.

Conclusions

Genetic diagnosis is a diverse field with a growing impact on public health in this country. Prenatal diagnosis, carrier screening, and newborn screening already play a major role in the management of inherited metabolic disorders. Advances in recombinant DNA technology and related fields will expand this role.

Diagnosis is only one answer to the problem of genetic diseases, however. A background paper on human gene therapy produced by the congressional Office of Technology Assessment (OTA) cites two major reasons that inherited diseases will never be eliminated, even if diagnostic tests could be developed for every known gene defect. The first is the constant appearance of new mutations. New mutations are responsible for a high proportion of many autosomal dominant disorders and for a smaller proportion of X-linked disorders.

The second reason concerns freedom of choice. The OTA report states that

> there are families for whom the prospect of selective abortion is unacceptable, or who choose not to avail themselves of genetic testing technologies for other ethical, religious, legal, social, or medical reasons. Such couples, while not at increased risk of having children with genetic diseases, will nevertheless inevitably bear some children with genetic defects. The only way to avoid this would be to circumscribe their liberty, making the judgment that the potential social benefit overrides their autonomous right to choose what is best for themselves and their families. The generally high regard for personal autonomy in our society implies that such couples' right to make reproductive decisions will be protected.[7]

Genetic diseases will continue to occur and new forms of therapy will be needed to treat them. Chapter 4 explores how the benefits and risks of gene therapy will compare with those of other treatment modalities.

―― ACKNOWLEDGMENTS ――

Chapter 3 is based in part on the presentation of Maurice J. Mahoney.

―― NOTES ――

1. A recent modification of standard restriction enzyme techniques, described by Stephen Embury and his coworkers in 1987 in

the *New England Journal of Medicine* 316(11):656–661, will allow prenatal diagnosis of sickle cell disease on the same day that the fetal DNA arrives in the laboratory.

2. Spina bifida and other neural tube defects are believed to be caused by a combination of heredity and environmental factors—they are multifactorial disorders.

3. DNA studies provide more precise information for some families than for others. In a 1987 article describing the use of DNA linkage analysis to identify carriers of DMD, Dr. C. Thomas Caskey says, "Carrier detection is highly accurate in families with multiple affected males. It is less accurate for females in those families with a single affected male since the origin of the mutation is often difficult to identify." *Science* 236:1225.

4. The safe administration of potent antiarrhythmic drugs to the fetus via the mother requires very careful monitoring of both the mother and the fetus. For more information, see Charles S. Kleinman et al., "In Utero Diagnosis and Treatment of Fetal Supraventricular Tachycardia," *Seminars in Perinatology* 9(2):113-129.

5. The findings are presented in a 1986 report by Frank Manning et al., *New England Journal of Medicine* 315(5):336-340.

6. The Johns Hopkins Hospital in Baltimore and the Massachusetts General Hospital in Boston began offering the test for Huntington disease in the fall of 1986. A recent article in the science magazine *Discover* notes that many of those who previously had said they would take advantage of such a test have not done so. Of 1,500 persons at risk for the disease in New England, only 32 had gone in for preliminary counseling by June 1987 (*Discover* 8:26-39).

7. *Human Gene Therapy—A Background Paper* (Washington, D.C.: U.S. Congress, Office of Technology Assessment, OTA-BP-BA-32, 1984), pp. 19-20.

—— SUGGESTED READINGS ——

Bickel, Horst, Robert Guthrie, and Gerhard Hammersen, eds. 1980. *Neonatal Screening for Inborn Errors of Metabolism.* New York: Springer-Verlag.

Blakeslee, Sandra. 1987. "Genetic Discoveries Raise Painful Questions." *New York Times,* April 21, pp. C1, C6.

Cahill, Kay. 1986 "Women Going Out of State for Prenatal Test." *Boston Globe,* December 1, p. 46.

Caskey, C. Thomas. 1987. "Disease Diagnosis by Recombinant DNA Methods." *Science,* June 5, 236:1223-1229.

Chervenak, Frank A., Glenn Isaacson, and Maurice J. Mahoney. 1986. "Current Concepts: Advances in the Diagnosis of Fetal Defects." *New England Journal of Medicine* 315(5):305-307.

Embury, Stephen H., Stephen J. Scharf, Randall K. Saiki, Mary Ann Gholson, Mitchell Golbus, Norman Arnheim, and Henry A. Erlich. 1987. "Rapid Prenatal Diagnosis of Sickle Cell Anemia by a New Method of DNA Analysis." *New England Journal of Medicine* 316(11):656-661.

Epstein, Charles J., David R. Cox, Steven A. Schonberg, and W. Allen Hogge. 1983. "Recent Developments in the Prenatal Diagnosis of Genetic Diseases and Birth Defects." *Annual Review of Genetics* 17:49-83.

Filkins, Karen, and Joseph F. Russo. 1985. *Human Prenatal Diagnosis.* New York: Marcel Dekker.

Genetic Counseling. 1985. Pamphlet produced by the March of Dimes Birth Defects Foundation.

The Genetic Resource. 1985. Special issue on the application of the maternal serum alpha-fetoprotein screening test. Published by the Massachusetts Genetics Program, Division of Family Health Services, Massachusetts Department of Public Health, Boston, Massachusetts. 2(2):8-29.

The Genetic Resource. 1986. Special issue on genetic screening. Published by the Massachusetts Genetics Program, Division of Family Health Services, Massachusetts Department of Public Health, Boston, Massachusetts. 3(1):1-56.

Grady, Denise. 1987. "The Ticking of a Time Bomb in the Genes." *Discover* 8(6):26-39.

Hamilton, Joan O'C., and Reginald Rhein, Jr. 1985. "The Giant Strides in Spotting Genetic Disorders Early." *Business Week,* November 18 pp. 82-85.

Human Gene Therapy—A Background Paper. 1984. Washington, D.C.: U.S. Congress, Office of Technology Assessment, OTA-BP-BA-32.

Kleinman, Charles S., Joshua A. Copel, Ellen M. Weinstein, Thomas V. Santulli, Jr., and John C. Hobbins. 1985. "In Utero Diagnosis and Treatment of Fetal Supraventricular Tachycardia." *Seminars in Perinatology* 9(2):113-129.

Knox, Richard A. 1986. "Genetic Tests Raise Concern about Abuse." *Boston Globe,* July 28, pp. 37-38.

Kolata, Gina. 1985. "Closing In on the Muscular Dystrophy Gene." *Science* 230:307-308.

——— 1986. "Two Disease-Causing Genes Found." *Science* 234:669-670.

Manning, Frank A., Michael R. Harrison, Charles Rodeck, and Members of the International Fetal Medicine and Surgery Society. 1986. "Special Report: Catheter Shunts for Fetal Hydronephrosis and Hydrocephalus." *New England Journal of Medicine* 315(5):336-340.

Modell, B., R. H. T. Ward, and D. V. I. Fairweather. 1980. "Effect of Introducing Antenatal Diagnosis on Reproductive Behavior of Families at Risk for Thalassaemia Major." *British Medical Journal* 280:1347-1350.

Nyhan, William L. 1985. "Neonatal Screening for Inherited Disease." *New England Journal of Medicine* 313(1):43-44.

Orkin, Stuart H. 1986. "Reverse Genetics and Human Disease." *Cell* 47:845-850.

Pang, Songya, and Maria I. New. 1985. "Past, Present, and Future Needs of Neonatal Screening in Congenital Disorders and in Inborn Errors of Metabolism." In Raul A. Wapnir, ed., *Congenital Metabolic Diseases: Diagnosis and Treatment*, pp. 85-102. New York: Marcel Dekker.

President's Commission for the Study of Ethical Problems in Medicine and Biomedical and Behavioral Research. 1983. *Screening and Counseling for Genetic Conditions*. Washington D.C.: U.S. Government Printing Office, Stock no. 040-000-00461-1.

Saltus, Richard. 1986. "Test Identifies Who Will Get Huntington's Disease." *Boston Globe,* October 3, p. 3.

Schmeck, Harold M., Jr. 1986. "Advances in Genetic Forecasts Increase Concerns." *New York Times,* August 19, pp. C1, C9.

——— 1987. "Burst of Discoveries Reveals Genetic Basis for Many Diseases." *New York Times,* March 31, pp. C1, C3.

Simmons, Kathryn. 1985. "Diagnostic Medicine Gains from DNA Probes." *JAMA* 253(1):16-18.

Squires, Sally. 1986. "The Ethics of Genetic Counseling." *Washington Post Health,* November 25, pp. 12-15.

4 | Current Treatments for Genetic Diseases

A national business magazine recently described a lonely nine-year-old girl who wanted a puppy more than anything else in the world.[1] The child could not have a pet because her immune system was crippled by the lack of an essential enzyme. To avoid life-threatening infections, she attended school via a speakerphone in her home. Once a month, she underwent lengthy blood transfusions to bolster her minimal defenses against disease.

The article noted that a bone marrow transplant might have corrected the girl's genetic defect, but that her parents had decided the procedure was too risky. Instead, they were "holding out for a startling new development called gene therapy."

This story illustrates the very difficult decisions that face parents who are forced by the birth of a sick child to become familiar with the complex field of medical genetics. The rapid pace of technological advances has created hope in situations that were once hopeless, but it also has led to a bewildering maze of choices—choices about when and where to seek treatment, choices about participation in early clinical trials of highly experimental procedures.

Decisions involving gene therapy are particularly difficult, because the potential benefits and risks associated with the technique are far from certain. In addition, no one knows for sure when gene therapy will become available. As stated in Chapter 1, many researchers expected the first clinical trials to take place before 1986, but that did not happen. For children who live with a life-threatening disease, each year of waiting represents an enormous challenge.

This chapter attempts to place gene therapy research in perspective by reviewing some of the existing therapeutic measures for genetic diseases, including metabolic manipulation (dietary therapy and related procedures), amplification of enzyme activity, replacement of a gene product, and organ and bone marrow transplantation. Such a review is important for two reasons. First, most researchers agree that, at least in the beginning, gene therapy should not be attempted for diseases that respond well to other forms of treatment.

The second reason is that scientists can learn a great deal about the best ways to use gene therapy by studying how target diseases respond to other therapeutic measures. Diseases that are candidates for early gene therapy trials must be at least partly reversible; that is, researchers must have some indication that insertion of a healthy gene into somatic cells will lead to clinical improvement. At present, the best evidence that a disease will respond to gene therapy is prior successful bone marrow transplantation.

Metabolic Manipulation

Most inherited metabolic diseases result from one of two factors: the build-up of a toxic substance or the absence of an essential product. Metabolic manipulations seek to overcome these problems by limiting the intake of potentially toxic substances, by depleting stores of these compounds before they damage delicate tissues, or by replacing missing products.

Dietary Restriction

The first successful therapy for an inborn error of metabolism was the phenylalanine-restricted diet developed in 1953 for phenylketonuria (PKU) by the European physician Horst Bickel and his colleagues. Phenylalanine, an amino acid, is a major constituent of most dietary proteins. In normal individuals, excess phenylalanine is converted to another amino acid, tyrosine, by the enzyme phenylalanine hydroxylase. Children born with PKU cannot make this conversion; without treatment, phenylalanine builds up in the body, damaging delicate brain cells and causing severe mental retardation.

The phenylalanine-restricted diet has two components: a strict limit on protein consumption and a dietary supplement to ensure that patients receive sufficient levels of other essential amino acids. Frequent blood tests help physicians adjust the diet as patients grow.

Other disorders that respond to dietary restrictions include galactosemia, hereditary fructose intolerance, lactose intolerance, methylmalonic acidemia, and the urea cycle defects (characterized by the appearance of toxic levels of ammonia in the blood). A more complete list appears in Table 4.1. None of the diets can reverse damage that occurs before birth. Thus, some children with galactosemia have learning disabilities, speech defects, and, in girls, ovarian abnormalities, despite early treatment.

Successful dietary therapy requires strict compliance, but managing the special diets in the home and educating very young children to monitor their own diets when they are away from home can be enormous tasks (see Table 4.2). Several research teams have found that PKU children have lowered self-esteem compared with their non-PKU siblings, in part because of tensions surrounding dietary management.

Maintaining the motivation necessary for strict adherence

TABLE 4.1. Examples of diseases treated with dietary restriction.

Disease	Substance restricted
Familial lipoprotein lipase deficiency	Neutral fats
Fructose intolerance	Fructose
Galactosemia and galactokinase deficiency	Galactose
Lactase deficiency	Lactose
Methylmalonic acidemia and propionic acidemia	Protein
Phenylketonuria	Phenylalanine
Refsum syndrome	Phytanic acid
Urea cycle disorders	Protein

SOURCE: Adapted, with permission, from Stanbury et al., *The Metabolic Basis of Inherited Disease*, 5th ed. (New York: McGraw-Hill, 1983), table 1-12.

TABLE 4.2. Sample diets of three children with phenylketonuria: total daily intake.[a]

Patient 1 6 years, 2 months weight 26.3 kg	Patient 2 10 years, 11 months weight 38.2 kg	Patient 3 15 years, 11 months weight 57.1 kg

Formula[b]
5 tbsp Lofenalac[c]
15 tbsp Phenylfree[d]
1¼ tsp tyrosine
  powder
	Formula
	22 tbsp Phenylfree
	3 tsp tyrosine powder
		Formula
		25 tbsp Phenylfree
		12 tbsp cooking oil
		4½ tsp tyrosine
		  powder

Food
3 chocolate chip
  cookies
27 grapes
1 small bag
  McDonald's french
  fries
1 medium banana
13½ tbsp white rice
1 tbsp catsup
	Food
	2 tbsp applesauce
	1 tbsp Hershey's
	  chocolate syrup
	1 slice white bread
	2 tsp grape jelly
	5 sugar wafer cookies
	2 tbsp raisins
	1 medium baked
	  potato
	2 tsp butter
	½ cup sherbet
	16 oz fruit drink
	8 oz Pepsi Cola
		Food
		1¼ cups Booberry
		  cereal
		1 slice
		  Weightwatcher's
		  bread
		2 tbsp jelly
		1 hot dog roll
		small leaf lettuce
		3 medium slices of
		  pickle
		8 medium mushrooms
		  (sauteed)
		15 medium potato
		  chips
		4 chocolate chip
		  cookies
		1 apple
		2 tbsp Hershey's syrup
		48 oz soda
		18 oz HiC
		2 popsicles
		sugar candies (a few)

Provides
460 mg phenylalanine
  (17.5/kg)
47.2 gm protein
  (1.79/kg)
1,523 Kcalories
  (58/kg)
	Provides
	340 mg phenylalanine
	  (8.90/kg)
	52.4 gm protein
	  (1.4/kg)
	1,835 Kcalories
	  (48/kg)
		Provides
		441 mg phenylalanine
		  (7.72/kg)
		63.3 gm protein
		  (1.1/kg)
		4,450 Kcalories
		  (78/kg)

a. Each child's diet is carefully calculated to meet his specific needs.

b. The formula is made up in a convenient volume and divided among meals.

c. Lofenalac is made by Mead Johnson & Co. It contains 7.5 mg phenylalanine per tablespoon (tbsp).

d. Phenylfree is made by Mead Johnson & Co. It contains no phenylalanine.

SOURCE: Dr. Mary Ampola, Amino Acid Laboratory, Boston Floating Hospital for Infants and Children, New England Medical Center, Boston, Massachusetts.

to dietary restrictions may be especially difficult for diseases such as PKU, in which the build-up of toxic substances does not cause acute physical symptoms. Initially, scientists believed that PKU children could make a safe transition to a normal diet shortly before they entered school (at age 4 1/2), but several studies have shown that some young children suffer an irreversible drop in IQ when they stop treatment. Today, some centers slowly decrease dietary controls for PKU between 8 and 10 years of age, while others encourage patients to stay on the diet through adolescence or later.

A new challenge to medical geneticists arose when the first group of patients treated with dietary therapy for PKU reached reproductive age. An international survey, reported by Boston researchers Roger R. Lenke and Harvey L. Levy in 1980, found that almost all babies born to untreated PKU mothers (mothers who had PKU but who were not on a phenylalanine-restricted diet during pregnancy) were mentally retarded, even though the babies did not have PKU. High phenylalanine levels in the intrauterine environment appear to prevent normal brain development in the fetus. Physicians recommend that women who have PKU begin dietary therapy before conception, or as early as possible in pregnancy. Unfortunately, some women who were treated as young children may not remember the reason for their special diet by the time they reach adulthood. Several medical centers have established programs to contact former PKU patients to warn them of the risk.

Depletion Techniques

Depletion techniques vary depending on the nature of the stored substance and the organs affected. In Wilson disease, excessive accumulation of copper in the liver and brain causes liver damage and a variety of neurological symptoms. Early diagnosis and removal of stored copper with the drug penicillamine prevents tissue damage and allows patients to lead normal lives. In idiopathic hemochromatosis (an autosomal recessive disease), massive iron overload damages the liver, heart, pancreas, endocrine glands, skin, and joints. Weekly bloodletting over a period of several years removes the accumulated iron, reverses heart and skin abnormalities, and arrests liver damage.

Physicians use several different mechanisms to prevent the build-up of excess uric acid in primary gout. One technique employs drugs such as probenecid to increase the excretion of uric acid by the kidneys. An alternative approach is to prevent the formation of uric acid by blocking the activity of the enzyme xanthine oxidase with the drug allopurinol.

Depletion techniques combined with dietary restriction also have been effective in familial hypercholesterolemia and in Refsum disease (in which neurological abnormalities, hearing loss, and night blindness result from the accumulation of phytanic acid in blood and tissues). Treatment for heterozygotes and homozygotes with familial hypercholesterolemia is directed at lowering the level of low density lipoprotein (LDL) in the blood. In heterozygotes, the most effective therapy is combined administration of a bile-acid-binding resin (which removes sterol from the body and enhances LDL receptor activity in the liver) and nicotinic acid. Homozygotes are resistant to drug therapy. Their blood levels can be lowered by (1) repeated plasmapheresis (in which blood is removed from the body and then centrifuged to separate blood cells from plasma, the fluid portion of the blood; the blood cells are returned to the patient and the plasma is discarded) or (2) a surgical procedure that allows part of the normal blood supply to bypass the liver. In patients with Refsum disease, periodic plasmapheresis (combined with dietary restriction of phytanic acid) may lead to improved muscle function and stabilization of vision problems.

Product Replacement

Hypothyroidism (cretinism), pituitary dwarfism, congenital adrenal hyperplasia syndrome, and orotic aciduria (associated with anemia, mental retardation, and growth retardation) result from enzyme defects that prevent the body from making an essential metabolic product. Replacement of these products (thyroid hormone for hypothyroidism, growth hormone for pituitary dwarfism, appropriate steroids for adrenal hyperplasia, and uridine for orotic aciduria) allows patients to lead normal or near-normal lives.

Until recently, the growth hormone used to treat pituitary dwarfism was extracted from human pituitaries obtained at autopsies. When several recipients of this hormone succumbed to a rare, infectious neurological disorder, the product was withdrawn from the market. Fortunately, human growth hormone made with recombinant DNA techniques became available several months later. The recombinant protein is safer because it is free of the risk of contamination with human viruses.

Amplification of Enzyme Activity

Many enzymes cannot function properly without the help of a vitamin-derived or metal cofactor. If the cofactor is not available (perhaps because of improper processing of a vitamin precursor), or if a genetic defect alters enzyme-cofactor binding, metabolic abnormalities may cause tissue damage. In some patients, the administration of large amounts of the appropriate cofactor can enhance residual enzyme activity and lead to clinical improvement. Several diseases, including methylmalonic acidemia and homocystinuria (which causes mental retardation, dislocation of the lenses of the eyes, skeletal abnormalities, and excessive blood clotting) have vitamin-responsive and nonresponsive subtypes.

Another form of enzyme amplification is possible in Crigler-Najjar syndrome type 2, which is caused by partial deficiency of a liver enzyme that breaks down the waste substance bilirubin. Patients have jaundice (characterized by yellowness of the skin, fatigue, and poor appetite) but are usually normal in other respects—though some have minor neurological abnormalities. Phenobarbital stimulates the production of the deficient enzyme sufficiently to correct the jaundice.

Replacement of the Gene Product

The ideal treatment for an inherited metabolic disorder, short of replacing the missing gene, is to replace the missing gene product. Gene product replacement has been very successful in the treatment of diseases associated with deficient blood proteins. It has been much less successful in the treatment of disorders involving intracellular enzymes (enzymes that act inside cells).

Blood Proteins

The most common genetic disease treated with product replacement is hemophilia. The availability of concentrated factor VIII (the missing blood clotting protein in hemophilia A) has produced a dramatic increase in the life expectancy of hemophiliacs. The major complications of this treatment are the development of immune inhibitors of the concentrate (which occurs in about 15 percent of severe hemophiliacs), hepatitis, and infection with HIV (human immunodeficiency virus, the virus that causes AIDS). Severe hemophiliacs may be exposed to as many as 100,000 blood donors per year because of the pooling process by which the concentrate is made. Recent efforts to screen blood for HIV and hepatitis viruses, combined with new virus inactivation procedures, have markedly reduced the risk of infection. However, between 60 and 80 percent of severe hemophiliacs now test positive for HIV antibodies because of exposures before 1984. (Scientists currently estimate that 25 to 50 percent of persons infected with HIV will progress to AIDS within 5 to 10 years of infection.)

Recombinant DNA technology may eliminate the need for human factor VIII concentrate. Both factor VIII and factor IX (the missing blood clotting protein in hemophilia B) have been produced in the laboratory by bacteria containing copies of the human genes. Researchers began clinical trials of recombinant factor VIII in 1987.

Another disorder of blood proteins that responds to gene product replacement is agammaglobulinemia (the absence of one or more types of antibody). Fortunately, intramuscular gamma globulin (the protein fraction of blood that contains antibodies) has not been implicated in the transmission of HIV.

Enzyme Replacement Therapy

The concept of enzyme replacement therapy was developed in the mid-1960s, when scientists discovered that they could "correct" abnormal storage patterns in skin cells taken from patients with certain lysosomal storage disorders (for example, Gaucher disease). These disorders result from the build-up of large carbohydrate or fat molecules in internal cellular compartments called lysosomes. Patients have a genetic defect that prevents them from making a particular lysosomal enzyme—without the enzyme, cells cannot degrade the large molecules into smaller molecules that could be reused or eliminated. The build-up of large carbohydrate or fat molecules causes brain and spinal cord abnormalities in some disorders, and in others produces kidney failure, liver disease, or cardiovascular problems.

Skin cells were able to rid themselves of accumulated chemicals when the appropriate enzyme was added to the culture medium in which the cells were grown. Researchers working with severely ill patients reasoned that the administration of missing enzymes might have the same effect inside the body. Initially, patients were treated with partially purified enzymes from nonhuman sources, because purified human enzymes were not available. Unfortunately, none of these trials produced clinical improvement, and patients often had severe adverse reactions to the foreign materials. Attempts to use human plasma or plasma concentrate as an enzyme source also failed.

The first clear indication that the administration of enzyme could reduce the quantity of certain stored substances in the body came from a 1971 study in which the enzyme hexosaminidase A (isolated from human urine) was given to a 12-month-old baby with Tay-Sachs disease. Researchers demonstrated that the enzyme was taken up rapidly by liver cells and that it degraded a substance called globoside. Unfortunately, this positive result was balanced by the finding that none of the external enzyme reached the baby's brain, the primary site of disease (accumulation of the sphingolipid ganglioside G_{M2} destroys nerve cells, leading to rapid degeneration of the nervous system and death).

Scientists were not surprised by this result, but they were disappointed. Researchers in many laboratories are now working on techniques to temporarily alter the physiological barrier between the bloodstream and the brain. Animal studies indicate, however, that even with barrier modification only about 1 percent of infused enzyme reaches the brain. Thus, enzyme replacement is not appropriate at this time for Tay-Sachs disease or for other conditions that affect the central nervous system.

Progress in enzyme replacement therapy for other lysosomal storage disorders has been more encouraging, although still painfully slow. The most extensive trials have been carried out in patients with Gaucher disease type 1 (in which the accumulation of fat molecules in cells called macrophages causes enlargement of the spleen and liver, bone damage, clotting problems, and anemia) and Fabry disease (in which patients may develop kidney damage and severe arteriosclerosis, leading to premature heart attacks and strokes).

Most of the clinical trials in Gaucher disease have used enzyme purified from human placenta. When this is injected into the body, it is removed rapidly from the bloodstream by liver cells (hepatocytes); very little reaches the affected macrophages.

In 1978, researchers demonstrated that the structure of the placental enzyme was responsible for this preferential uptake by liver cells. Most lysosomal enzymes are glycoproteins; they consist of a protein chain chemically bonded to a carbohydrate (sugar) structure. The significant finding in Gaucher disease was that a portion of the sugar structure on the placental enzyme was acting as an address label for surface receptors on the liver cells. (Surface receptors on cells and tissues are required for the uptake of many substances from the bloodstream; each type of receptor recognizes a specific chemical structure.) To overcome this problem, scientists have devised biochemical techniques to remove some of the sugar residues from the enzyme. This process exposes structures that are more likely to be recognized by receptors on the abnormal macrophages.

Special cellular uptake systems are being investigated for several lysosomal storage disorders. Researchers study target cells (cells that are being damaged in a particular disease) to identify and characterize their surface receptors. Then they attempt to alter the appropriate purified enzyme to enhance its uptake by those receptors.

Another approach to delivering enzymes is to use biodegradeable carrier vesicles. This allows larger quantities of enzyme to reach target tissues. Researchers have entrapped enzymes in vesicles formed from red blood cell membranes and in man-made microscopic spheres called liposomes. Theoretically, the surfaces of both types of vesicles could be modified to enhance their uptake by cell-specific receptors.

The red blood cell membranes appear to protect enzymes from degradation in the bloodstream and from destruction by the immune system. These artificial cells have been used experimentally in patients with Gaucher disease. The studies demonstrated that the vesicles are safe, but patients have not yet shown signs of clinical improvement.

In contrast, animal studies indicate that liposomes may not be a suitable tool in the treatment of genetic diseases. Study animals have experienced severe immune reactions and other dangerous side effects from the liposome-enzyme complexes.

Laboratory and clinical studies suggest that in some disorders caused by deficient or absent enzyme activity, it might not be necessary to deliver the normal enzyme directly to target cells. These are disorders in which the substance that causes tissue damage flows freely in and out of affected cells. In such cases, researchers might be able to relieve the buildup of the harmful substance by keeping large quantities of active enzyme in the bloodstream. This task is complicated, however, because of the rate at which purified enzymes are removed from the circulation. The degradation of a foreign enzyme occurs quite rapidly, especially if it elicits an immune response in the recipient.

Many strategies have been explored to overcome this

problem. The most successful was described recently by Michael S. Hershfield of Duke University Medical Center and his coworkers in the *New England Journal of Medicine* (March 5, 1987). Although the researchers were seeking a new treatment for adenosine deaminase (ADA) deficiency, their results may signal a new direction in enzyme replacement therapy for many genetic disorders.

The treatment developed by Dr. Hershfield and his collaborators employs polyethylene glycol (PEG), an inert waxy material. The researchers chemically linked PEG to purified ADA obtained from cows. They found that the waxy substance formed a protective barrier that prevented the degradation of ADA by enzymes normally found in the bloodstream. In early studies in mice, PEG also blocked the development of a specific immune response to the foreign protein. In fact, the attachment of PEG to ADA increased the enzyme's half-life—the amount of time required for half of the ADA to be removed from the circulation in the mouse—from a few minutes to 24 hours.

The Duke University scientists have administered weekly intramuscular injections of PEG-ADA to two patients with severe combined immune deficiency. The first is a four-year-old girl. Previously, the child had failed to respond to two bone marrow transplants from her father or to periodic red blood cell transfusions. After two months on the PEG-ADA regimen, her immune functions began to improve. At the end of six months, her responses approached the normal range on several tests of lymphocyte activity (see Figure 4.1). She has gained weight and shows no signs of the recurrent infections that plagued her before therapy began. She has had no adverse reactions to the injections. Similar positive results have been observed in a ten-year-old girl who was beginning to show signs of iron overload from prolonged transfusion therapy.

The researchers caution that preliminary results in two patients are not sufficient to prove the efficacy and safety of PEG-ADA, but they believe the short-term findings are very encouraging. They suggest that PEG-modified enzymes might help patients with other disorders in which toxic products inside the cell are in equilibrium with those in the bloodstream.

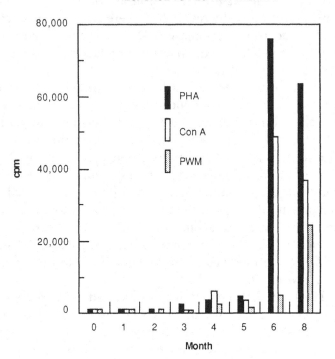

FIGURE 4.1 This graph shows one effect of PEG-ADA therapy on the immune function of a patient with ADA deficiency. In laboratory studies, researchers examined the response of the patient's blood lymphocytes to three substances that stimulate the proliferation of immune cells in a normal individual: phytohemagglutinin (**PHA**), concanavalin A (**Con A**), and poke-weed mitogen (**PWM**). Every month, lymphocytes were isolated from the blood and incubated in the presence of the stimulating substances for several days. The number of actively dividing cells after the incubation period was measured by the ability of the cells to incorporate radioactively labeled thymidine (the amount of thymidine incorporated is given in cpm, or counts per minute). These tests provide an indication of the ability of the immune cells to respond to foreign substances. Before treatment and during the first two months of therapy, there was almost no response to stimulation. By the sixth month, the level of response had reached close to normal levels for the three stimulatory agents. SOURCE: From M. S. Hershfield et al., "Treatment of Adenosine Deaminase Deficiency with Polyethylene Glycol-Modified Adenosine Deaminase," *New England Journal of Medicine* 316(10):589-596. Reprinted by permission of The New England Journal of Medicine.

Transplantation Therapy

Metabolic manipulations and gene product replacement both require frequent repetitions of therapeutic measures to protect affected children from life-threatening deterioration of their condition. While these techniques represent a vast improvement over the helplessness of past generations in coping with genetic disease, they are far from ideal solutions. To effect a cure, researchers must be able to provide patients with the ability to make normal gene products on their own.

Organ transplantation is the one strategy now available that accomplishes this goal. Organs from healthy donors continue to make normal enzymes when they are transplanted into patients with specific genetic defects. Recent improvements in the management of graft rejection and other complications have made transplantation a feasible alternative for a growing number of inherited disorders (see Table 4.3).

The decision to perform a transplant depends on the severity of the patient's condition, the risks of the procedure, and the expected benefits. Most transplantation operations for genetic conditions involve disorders in which the clinical

TABLE 4.3. Examples of diseases in which transplantation has led to clinical improvement.

Disease	Transplanted organ
Alpha$_1$-antitrypsin deficiency[a]	Liver
Cardiomyopathy	Heart
Cystinosis	Kidney
Hemophilia A	Liver
Mucopolysaccharidoses	Bone marrow
Severe combined immune deficiency	Bone marrow
Thalassemia	Bone marrow
Tyrosinemia	Liver
Wilson disease	Liver

a. Patients with alpha$_1$-antitrypsin deficiency who escape liver disease in infancy usually develop lung disease as young adults. Physicians do not yet know whether liver transplant recipients will be protected from lung problems as they mature.

SOURCE: Robertson Parkman, M.D., Head, Division of Research Immunology and Bone Marrow Transplantation, Childrens Hospital of Los Angeles.

symptoms are localized, or limited to a single organ system. Generalized diseases—those affecting multiple organ systems—may be improved by transplantation if the normal enzyme or protein produced by the donor organ can gain access to the sites of clinical disease.

Solid Organs

Because of risks inherent in the transplantation procedures, solid organ transplantation usually is reserved for patients with imminent organ failure. Among these risks are rejection and the need for lifelong suppression of the immune system.

Liver. The solid organ most often transplanted in patients with inherited metabolic disorders is the liver. Localized diseases that have responded well to liver transplantation in some patients include tyrosinemia (8 cases),[2] glycogen storage disease type 4 (Andersen disease; 3 cases), and familial hypercholesterolemia (2 cases). Generalized diseases that have responded well to liver transplantation include alpha$_1$-antitrypsin deficiency (more than 60 cases), Wilson disease (17 cases), and hemophilia A (3 cases).

Patients with alpha$_1$-antitrypsin deficiency who escape liver disease in infancy usually develop lung disease as young adults; physicians do not yet know whether liver transplant recipients will be protected from lung problems as they mature.

Most patients with Wilson disease (in which the toxic accumulation of copper in the liver "overflows" into the central nervous system, causing involuntary movements and loss of coordination) respond well to the drug penicillamine; however, the lag time between the start of treatment and improvements in liver function may be three to six months. Patients whose Wilson disease is first diagnosed because of acute liver failure may not be able to wait that long. Successful liver transplantation in these people leads to normal liver function and reduces or eliminates neurological symptoms.

Dr. Thomas Starzl, professor of surgery and chief of the organ transplantation program at the University of Pittsburgh, notes that in several cases liver transplantation has provided important clues about the basis of an inherited disease. For example, the first liver transplantation in a patient with hemophilia A demonstrated that a crucial segment of the large factor VIII complex, which is inactive or missing in hemophiliacs, is produced by the liver.

The graft survival rate in liver transplantation for inborn errors of metabolism is about 80 percent after one year.

Heart. Between 60 and 70 patients with an inherited form of cardiomyopathy (an abnormality of the heart muscle) have received heart transplants. These patients do as well as other heart transplant recipients; the one-year graft survival rate is 85 percent and the five-year graft survival rate is 66 percent. Young patients may eventually have to receive a second transplant because of their increased incidence of coronary artery disease.

Kidney. Kidney transplantation produces mixed results in patients with inherited metabolic disorders. Transplantation has been used to treat kidney failure in Alport syndrome, cystinosis, Fabry disease, amyloidosis, oxalosis, and several other conditions. In Alport syndrome, cystinosis, and Fabry disease, the transplanted normal kidney is not affected by the underlying disease. In contrast, oxalosis (characterized by the development of kidney stones and calcium deposits) recurs so rapidly in the new organ that transplantation is no longer recommended for oxalosis patients. The one-year survival rate for kidney grafts used to correct a genetic defect is between 80 and 90 percent. The five-year graft survival rate is between 50 and 70 percent.

Initially, researchers hoped that kidney transplantation would do more than restore kidney function in patients with genetic diseases. They regarded the procedure as a form of enzyme replacement therapy. Normal enzymes from the graft were expected to reduce toxic accumulations throughout the body (either the normal kidney would take up and metabolize the unwanted substance from the circulation, or enzymes released by the normal kidney would be taken up by affected tissues to relieve their storage problems). Attempts to document such an effect have been inconclusive.

Bone Marrow Transplantation

Bone marrow transplantation offers a more versatile technique for replacing missing or defective enzymes than solid organ grafts, in part because the descendants of bone marrow cells migrate throughout the body.

Bone marrow can be viewed as two separate organs. The first organ is composed of lymphoid progenitor cells, which give rise to two basic elements of the immune system, T-lymphocytes (the cells responsible for tissue graft rejection and other aspects of cell-mediated immunity) and B-lymphocytes (antibody-producing cells). The second consists of hematopoietic progenitor cells, which give rise to red blood cells, platelets, osteoclasts (cells associated with bone growth), and cells of the mononuclear phagocyte system. (Phagocytes are scavenger cells; they ingest cellular debris, bacteria, and other foreign particles. Cells of the mononuclear phagocyte system include monocytes in the bloodstream, Kupffer cells in the liver, alveolar macrophages in the lung, and, possibly, microglial cells in the brain.)[3]

For some patients, bone marrow transplantation can provide complete freedom from the limitations of a severe genetic disease, but the benefits must be weighed carefully against the risks in each case. All forms of transplantation carry the risk that immune cells in the recipient will recognize donor cells as foreign and destroy them, resulting in rejection of the graft. Bone marrow recipients face the additional problem of **graft versus host disease,** or GVHD. In GVHD, immune cells derived from the donor bone marrow recognize the host cells as foreign and attack them, injuring the skin, the liver, the intestinal tract, and other tissues. Between 10 and 20 percent of bone marrow transplant patients succumb to GVHD and related complications within a year of the procedure.

To reduce the likelihood of a life-threatening immune reaction, bone marrow transplantation for genetic diseases usually is limited to patients who have genetically matched siblings. The term "matched" refers to cell-surface proteins called **histocompatibility antigens.** These proteins constitute a recognition system that enables immune cells to distinguish between "self" and "nonself." The probability that any two siblings are histocompatible is 1 in 4. Only about 30 percent of patients with treatable diseases have a histocompatible donor.

The likelihood of a bone marrow transplant curing or stabilizing a patient with a particular genetic disease depends on (1) the tissue in which the normal gene product is expressed; (2) the spectrum of clinical symptoms associated with the gene defect; and (3) the transport mechanisms that exist for the product within affected cells. Diseases involving gene products not expressed by bone marrow cells, including cystic fibrosis, hemophilia, and phenylketonuria, will not respond to bone marrow transplantation.

Diseases in Bone Marrow-Derived Cells. Histocompatible bone marrow transplantation is the treatment of choice for severe combined immune deficiency (SCID) and Wiskott-Aldrich syndrome (a disorder that affects both lymphoid and hematopoietic cells). It also has been employed successfully in thalassemia, severe type Gaucher disease, sickle cell disease, osteopetrosis (which leads to bone deformities, neurological abnormalities, and facial paralysis), and many different phagocytic disorders (see Table 4.4).

The first successful bone marrow transplantation was performed in a child with SCID in 1968. SCID patients are particularly good candidates for grafting because their immune defect makes them less likely to reject the donor's transplanted bone marrow cells.[4] Following transplantation with histocompatible normal bone marrow (containing both lymphoid and hematopoietic progenitor cells), 90 percent of SCID patients produce lymphocytes of donor origin. However, red blood cells, phagocytic cells, and all other blood components retain the genetic makeup of the host.

The different fates of lymphoid and hematopoietic progenitor cells in these patients illustrate the very important concept of "physiological space." Successful engraftment requires that donor cells compete effectively for space in the host environment. In SCID, lymphoid progenitor cells from the donor grow and multiply because the genetic defect puts the recipient's own cells at a competitive disadvantage. However, the recipient's hematopoietic progenitor cells function normally, so there is no physiological space for the foreign hematopoietic progenitor cells, and they eventually disappear.

Most genetic diseases do not affect the viability of lymphoid or hematopoietic progenitor cells. Therefore, physicians have to create physiological space for transplanted cells by destroying the recipient's existing bone marrow before injecting donor cells. This is accomplished with highly toxic chemotherapeutic drugs, alone or in combination with radiation. The most common preparative regimen employs the drugs cyclophosphamide to eliminate lymphoid progenitor cells and busulfan to eliminate hematopoietic progenitor cells. Once these drugs have been administered, the patient is extremely

TABLE 4.4. Genetic diseases for which bone marrow transplantation has been attempted.

Disease cured or stabilized	Limited improvement[a]	No improvement
Chediak-Higashi syndrome	Adrenoleukodystrophy	Lesch-Nyhan syndrome
Chronic granulomatous disease	Hunter syndrome	Pompe disease
Fanconi anemia	Hurler syndrome	
Gaucher disease	Metachromatic	
Granulocyte actin deficiency	leukodystrophy	
Infantile agranulocytosis	Maroteaux-Lamy	
Osteopetrosis	syndrome	
Purine nucleoside phosphorylase deficiency	Sanfilippo syndrome	
Severe combined immune deficiency		
Sickle cell disease		
Thalassemia		
Wiskott-Aldrich syndrome		

a. Long-term impact of bone marrow transplantation on symptoms affecting the central nervous system remains unclear.

SOURCE: Robertson Parkman, M.D., Head, Division of Research Immunology and Bone Marrow Transplantation, Childrens Hospital of Los Angeles.

susceptible to infection until donor cells have repopulated the bone marrow. (GVHD may delay engraftment, increasing the period of extreme susceptibility to infection.) Long-term side effects of these preparative procedures have not been fully characterized, but it is known that most patients become sterile.

Experience with disorders affecting bone marrow-derived cells indicates that bone marrow transplantation is more successful if it is performed early in the course of a disease rather than later. One reason for this is that GVHD is less common in young children than in older people. Also, young children are less likely to be carriers of viral infections (especially cytomegalovirus) that could become fatal in the absence of normal immune function. In addition, older patients who have had multiple transfusions—for thalassemia or very severe sickle cell disease—are more likely to reject a graft than younger patients, because their T lymphocytes have become highly sensitized to foreign tissues.

Another consideration in the decision about when to employ bone marrow transplantation is the slow turnover of bone marrow-derived cells in tissue. In one patient with Gaucher disease, enzyme activity in circulating white blood cells reached normal levels within one month of transplantation; but abnormal Gaucher cells did not begin disappearing from the bone marrow until four months later. Similarly, the improvement of patients with osteopetrosis starts three to four months after transplantation.

Generalized Genetic Diseases. The impact of bone marrow transplantation on diseases affecting multiple tissues depends (1) on the relative amount of enzyme produced by bone marrow-derived cells compared with that normally produced in the body; (2) on the ability of affected cells to actively take in enzyme from their environment; and (3) on the accessibility of affected tissues, particularly the central nervous system.

Bone marrow transplantation has reversed spleen and liver enlargement and corrected other physical abnormalities in patients with several of the mucopolysaccharidoses (lysosomal storage disorders involving the abnormal accumulation of mucopolysaccharides). In a patient with severe Maroteaux-Lamy syndrome, bone marrow transplantation resulted in a dramatic improvement in cardiac and pulmonary function, decreased liver and spleen size, and slightly improved joint mobility. However, transplantation did not alter the patient's severe bony disease or corneal clouding.

Several diseases have not responded to bone marrow transplantation. Among them is Pompe disease, caused by a deficiency of the enzyme alpha-1,4-glucosidase. Without this enzyme, glycogen accumulates in the lysosomes of the muscle and heart, resulting in muscle weakness and heart failure. Following transplantation, levels of the enzyme in the blood become normal, but affected muscle and liver cells retain their abnormal glycogen deposits.

Disorders of the Central Nervous System. Enzyme replacement studies and early trials in laboratory animals indicated that bone marrow transplantation might not have an effect on storage abnormalities in the brain, because it appeared that enzymes produced by bone marrow-derived cells could not cross the blood–brain barrier. However, recent clinical reports provide grounds for cautious optimism.

William Krivit of the University of Minnesota Medical School in Minneapolis notes that the most encouraging data concern a bone marrow transplant in a Minnesota patient with metachromatic leukodystrophy (MLD, an autosomal recessive disorder primarily affecting the central nervous system). The girl's sister had progressed from the onset of symptoms to total loss of motor control and inability to swallow in less than a year. In contrast, the treated patient, who received a bone marrow transplant as soon as she began to show signs of the disease, was achieving appropriate milestones in development (including language, motor, and social skills) 15 months later.

Similarly, physicians at the Westminster Children's Hospital in London report stabilization or improvement of psychomotor development in five children who received bone marrow transplants for Hurler syndrome (usually associated with progressive, severe mental retardation).

One possible explanation for these results is that bone marrow-derived cells cross the blood–brain barrier and provide a local source of enzyme in the central nervous system. Physicians treating another patient with MLD found white blood cells of donor origin in the child's spinal fluid six months after bone marrow transplantation. The appearance of the cells corresponded with a change in the patient's clinical condition: his psychomotor condition stabilized.

This finding and recent results in studies involving mice and dogs underscore the importance of early intervention. The turnover of bone marrow-derived cells in the central nervous system appears to be very slow. During the months that the turnover is taking place, irreversible damage to nerve cells continues. In the twitcher mouse, a model for the human disorder Krabbe disease (in which death usually occurs before age two), bone marrow transplantation improves survival only if it is performed before affected mice are 10 days old.

The ultimate impact of bone marrow transplantation on disorders of the central nervous system is not yet known. In some disorders, bone marrow transplantation shortly after birth may prevent disease. In others, enzyme from bone marrow-derived cells may not be effective, or it may not be possible to achieve clinically significant concentrations of the normal enzyme in the central nervous system before irreversible brain damage occurs.

Conclusion

The results of histocompatible bone marrow transplantation provide important information about the potential success of gene therapy, because both procedures work by providing patients with bone marrow cells that express normal genes. In fact, given the similar outcomes, it might seem that gene therapy increases the risks without altering the benefits. However, bone marrow transplantation has three important limitations:

- Only about 30 percent of individuals have a histocompatible donor.

- Graft versus host disease can be a serious problem, even among patients who have acceptable donors.
- Potential recipients must be prepared with high doses of chemotherapy to reduce the risk of graft rejection and to provide physiological space for the donor cells.

In contrast, all patients have a potential donor for gene therapy—themselves. Also, the use of a person's own bone marrow would eliminate the risk of GVHD. The third issue is somewhat more complicated. Researchers suspect that some form of preparative chemotherapy may still be necessary in gene therapy to give genetically altered bone marrow cells room to grow and multiply.

Patients *without* histocompatible donors who have diseases that have been shown to respond to bone marrow transplantation are the most likely candidates for early gene therapy trials. As noted in Chapter 2, many researchers believe that the first trials will be done in patients with SCID. However, the recent success of enzyme replacement therapy with PEG-ADA may reduce the need for gene therapy in that particular group of patients.

—— ACKNOWLEDGMENTS ——

Chapter 4 is based in part on the presentations of Robertson Parkman and James Wyngaarden.

—— NOTES ——

1. *Business Week*, November 18, 1985, p. 76
2. Figures on the numbers of transplant procedures performed for specific disorders were presented by Dr. Robertson Parkman at the 1986 Annual Meeting of the Institute of Medicine.
3. Some researchers believe that microglial cells are derived from hematopoietic tissue, while others believe they are of neuroectodermal origin.
4. SCID is the one genetic disease for which physicians will use partially matched donors, often parents, if no histocompatible donor is available. T lymphocytes are removed from the donor marrow to reduce the risk of GVHD. The success rate of these transplants is

lower than that of histocompatible transplants (60 percent, compared with about 80 percent).

—— SUGGESTED READINGS ——

Bickel, Horst. 1985. "Differential Diagnosis and Treatment of Hyperphenylalaninaemia." In Kare Berg, ed., *Medical Genetics: Past, Present, Future*, pp. 93-107. New York: Alan R. Liss.

Brady, Roscoe O. 1985. "Lipid Storage Disorders Enzyme Replacement Therapy: Current Status and Future Strategies." In Raul A. Wapnir, ed., *Congenital Metabolic Diseases: Diagnosis and Treatment*, pp. 285-297. New York: Marcel Dekker.

Desnick, Robert J. 1987. "Bone Marrow Transplantation in Genetic Diseases: A Status Report." In Michael M. Kaback and Larry J. Shapiro, eds., *Frontiers in Genetic Medicine*, pp. 71-83. Report of the 92nd Ross Conference on Pediatric Research, June 1986. Columbus, OH: Ross Laboratories.

Desnick, Robert J., and Gregory A. Grabowski. 1981. "Advances in the Treatment of Inherited Metabolic Diseases." In Harry Harris and Kurt Hirschhorn, eds., *Advances in Human Genetics*, vol. 11, pp. 281-369. New York: Plenum Press.

Hershfield, Michael S., Rebecca H. Buckley, Michael L. Greenberg, Alton L. Melton, Richard Schiff, Christine Hatem, Joanne Kurtzberg, M. Louise Markert, Roger Kobayashi, Ai Lan Kobayashi, and Abraham Abuchowski. 1987. "Treatment of Adenosine Deaminase Deficiency with Polyethylene Glycol-Modified Adenosine Deaminase." *New England Journal of Medicine* 316(10):589-596.

Hugh-Jones, K. 1986. "Psychomotor Development of Children with Mucopolysaccharidosis Type 1-H Following Bone Marrow Transplantation." In William Krivit and Natalie W. Paul, eds., *Bone Marrow Transplantation for Treatment of Lysosomal Storage Diseases*, pp. 25-29. March of Dimes Birth Defects Foundation, Birth Defects Original Article Series, vol. 22 no. 1. New York: Alan R. Liss.

Krivit, William. "Conclusions: Ethics, Cost, and Future of Bone Marrow Transplantation for Lysosomal Storage Diseases." 1986. In William Krivit and Natalie W. Paul, eds., *Bone Marrow Transplantation for Lysosomal Storage Diseases*, pp. 189-194. March of Dimes Birth Defects Foundation, Birth Defects Original Article Series, vol. 22, no. 1. New York: Alan R. Liss.

Lenke, Roger R., and Harvey L. Levy. 1980. "Maternal Phenylke-tonuria and Hyperphenylalaninemia." *New England Journal of Medicine* 303(21):1202-1208.

Lewis, Jessica H., Franklin A. Bontempo, Joel A. Spero, Margaret V. Ragni, and Thomas E. Starzl. 1985. "Liver Transplantation in a Hemophiliac." *New England Journal of Medicine* 312(18):1189-1190.

Moen, Joan L., Robert D. Wilcox, and Joan K. Burns. 1977. "PKU as a Factor in the Development of Self-Esteem." *Journal of Pediatrics* 90(6):1027-1029.

O'Reilly, Richard J., Joel Brochstein, Robert Dinsmore, and Dahlia Kirkpatrick. 1984. "Bone Marrow Transplantation for Con-genital Disorders." *Seminars in Hematology* 21:188-221.

Parkman, Robertson. 1986. "The Application of Bone Marrow Transplantation to the Treatment of Genetic Diseases." *Science* 232:1373-1378.

Pentchev, Peter G., Roscoe O. Brady, Andrew E. Gal, and Sue R. Hibbert. 1975. "Replacement Therapy for Inherited Enzyme Deficiency: Sustained Clearance of Accumulated Glucocerebro-side in Gaucher's Disease Following Infusion of Purified Glu-cocerebrosidase." *Journal of Molecular Medicine* 1:73-78.

Schulman, Roger. 1985. "The Gene Doctors." *Business Week*, No-vember 18, pp. 76-80.

Stanbury, John B., James B. Wyngaarden, Donald S. Fredrickson, Joseph L. Goldstein, and Michael S. Brown. 1983. *The Meta-bolic Basis of Inherited Disease*, 5th edition. New York. McGraw-Hill.

Taylor R. M., G. J. Stewart, B. R. H. Farrow, and P. J. Healy. 1986. "Enzyme Replacement in Nervous Tissue after Allogeneic Bone-Marrow Transplantation for Fucosidosis in Dogs." *Lancet* II:772-774.

Thomas, E. Donnall. 1985. "Marrow Transplantation for Nonma-lignant Disorders." *New England Journal of Medicine* 312(1):46-47.

5 | Preparation for Gene Therapy

Excitement over the prospect of human gene therapy was at its height in 1985. Physicians saw gene therapy as a way of extending the benefits of bone marrow transplantation to patients with genetic diseases who had no compatible donors. Reports from the laboratory looked encouraging. The first genetic "cure" in mammals had been described a year earlier.[1] Also, researchers in Canada and the United States had developed efficient new techniques for inserting foreign DNA into bone marrow cells. In November 1985, W. French Anderson of the National Institutes of Health wrote, "Before 1986 one or more physicians probably will have submitted clinical protocols [to local and national review boards] requesting permission to carry out this procedure in humans."[2]

Almost a year later, at the October 1986 meeting of the Institute of Medicine, Dr. Anderson and others acknowledged that their early predictions were overly optimistic. Long-term experiments in mice and isolated studies in larger mammals have raised new questions about the efficacy and reliability of current gene transfer methods. A gene that is inserted into a cell rarely functions as well (produces as much protein) as an identical gene in its natural environment. Also, cells that are carrying foreign genes sometimes behave differently in the test tube than they do inside a living animal. Scientists in many laboratories are working to understand these problems and to develop solutions, but few expect human trials to begin before 1989.

The basic tasks in gene therapy are:

(1) isolating the gene responsible for a genetic disease and producing multiple copies of its normal counterpart;
(2) selecting target cells in which to deliver the gene to the body;
(3) inserting the normal gene into the target cells;
(4) achieving appropriate expression of the inserted gene (that is, production of an appropriate amount of protein).

This chapter describes various approaches to the first three tasks. The following chapter addresses the issue of expression and examines the important role of animal models in the development of new gene transfer techniques.

These two chapters contain sufficient detail for those readers who wish to acquire a basic understanding of the molecular biology of gene therapy. Readers who do not wish to dwell on the intricacies of the technical achievements will find that the conclusions at the end of each chapter and the summary in Chapter 9 provide an overview of recent progress in the field and of the challenges that lie ahead.

Finding and Copying the Gene

Scientists must know the identity of the gene responsible for a genetic disease before they can begin to consider gene therapy. The task of picking one defective gene out of the approximately 100,000 genes that constitute the human genome may seem overwhelming, but recent advances in molecular biology have made the process relatively straightforward. The choice of an appropriate technique depends, in large part, on whether scientists know the identity of the abnormal gene's protein product.

If the Gene Product is Known

If the abnormal protein has been identified, researchers can trace the problem back to the genome in two different ways.

The first involves a precise analysis of the protein's molecular structure—specifically, the sequence of amino acid building blocks in the protein. The second employs an intermediary in the protein-building process called **ribonucleic acid,** or RNA.

Protein Structure.   As explained in Chapter 2, the sequence of amino acids in a protein molecule is determined by the sequence of nucleotide triplets, or codons, in the corresponding gene. Researchers searching for a defective gene work backwards; using their knowledge of the genetic code, they synthesize a small DNA molecule based on part of the structure of the defective protein.

The synthetic DNA molecule will stick to any matching nucleotide sequence in the human genome. Thus, researchers can use the synthetic molecule as a probe to locate the gene of interest. (Probes are tagged with radioactive tracers to make them stand out in the midst of large amounts of DNA.)

RNA Structure.   Most of the DNA in an animal cell is located in a spheroidal body called the **nucleus.** A type of RNA called **messenger RNA** (mRNA) carries genetic instructions from the nucleus to the cytoplasm, the site of cellular machinery required to make proteins. When a cell needs a new protein molecule for a particular function or structure, the portion of the DNA molecule coding for that protein is copied into mRNA in a process called **transcription** (each DNA triplet codon is transcribed into a corresponding RNA triplet codon). The mRNA then migrates to the cytoplasm, where each mRNA codon is **translated** into a protein building block.

In some cases, researchers can isolate and purify the mRNA that codes for an abnormal protein. The mRNA molecule is then used as a template to synthesize a complementary piece of DNA (called cDNA). When the cDNA is labeled with a radioactive tracer, it becomes a probe to identify the abnormal gene. The cDNA also can be cloned (see below).

If the Gene Product is Not Known

Identifying the gene associated with a genetic disease is much more difficult if its protein product is not known, but researchers have made major strides in this area as well. In some cases, the discovery of chromosomal deletions in patients with a particular genetic disease has helped researchers narrow the search for an abnormal gene. The locations of other genes have been determined solely by linkage analysis techniques, described in Chapter 3.

Chromosomal Deletions. Using chromosomal deletions as a guide, different groups of scientists recently isolated the gene responsible for retinoblastoma (an inherited form of eye cancer that affects the retina, the site of visual receptors in the eye);[3] located a portion of the gene that causes Duchenne muscular dystrophy, one of the most common X-linked genetic diseases; and identified the gene associated with chronic granulomatous disease (CGD; an X-linked disorder in which patients have repeated, severe infections).

The investigators studying retinoblastoma found that many patients had deletions in a particular region of chromosome 13. This information allowed them to employ a mapping technique called "walking the genome." (After establishing a reference point, researchers use restriction enzymes and DNA probes to move along a chromosome, studying one small fragment of DNA at a time.) Eventually, a group headed by Thaddeus Dryja of the Massachusetts Eye and Ear Infirmary and by Stephen Friend and Robert Weinberg of the Whitehead Institute identified a fragment of chromosome 13 that appeared to satisfy the criteria for a retinoblastoma gene. Retinal cells in normal persons contain mRNA that matches the isolated DNA fragment, while cells from retinoblastoma patients do not.

Researchers looking for the muscular dystrophy gene had an advantage, because they knew the gene was on the X chromosome. Studies of chromosomal deletions in muscular dystrophy patients narrowed the search further to a region of the short-arm of the X chromosome (Xp21). Louis Kunkel of Children's Hospital in Boston and his associates identified part of the gene in 1986. Isolation of the remainder will allow scientists to pinpoint the protein abnormality that causes this wasting disease. Already, they have learned that the muscular dystrophy gene codes for a protein that is expressed only in muscle cells.

Two of the DMD patients with deletions in the Xp21 region of the X chromosome also had chronic granulomatous disease. Further studies of these patients, combined with linkage analysis studies of patients with classical X-linked CGD (not associated with DMD), enabled researchers at Boston Children's Hospital and Harvard Medical School to establish the exact location of the CGD gene. Stuart Orkin, one of the leaders of the Boston group, says that they have now taken the process one step further. In early 1987, they identified the gene's protein product. These accomplishments have made CGD a potential candidate for early gene therapy trials.

Linkage Analysis Techniques.   Chapter 3 describes the use of linkage analysis techniques in the study of Huntington disease. By comparing DNA fragments from patients with Huntington disease with DNA fragments from their healthy relatives (and treating both sets of fragments with the same radioactively labeled DNA probes), researchers have traced the gene that causes Huntington disease to chromosome 4. Efforts are under way to determine the gene's exact location and to isolate it from the surrounding DNA.

Gene Cloning

After a defective gene has been located, researchers must isolate and make multiple copies of its normal counterpart. This process, called **gene cloning** (see Figure 5.1), employs rapidly dividing single-celled organisms, such as bacteria or yeast,[4] to reproduce pieces of DNA from higher organisms.

The first step in cloning a human gene, once it has been identified, is to insert it into the single-celled organisms that will copy it. This is accomplished by combining the gene with a cloning "vehicle." The most common cloning vehicles are **plasmids**—small circular molecules of DNA that occur naturally in many different types of bacteria. They remain separate from the single bacterial chromosome, and they contain genes that are not essential for bacterial growth. Plasmids have the

MAJOR STEPS IN GENE CLONING

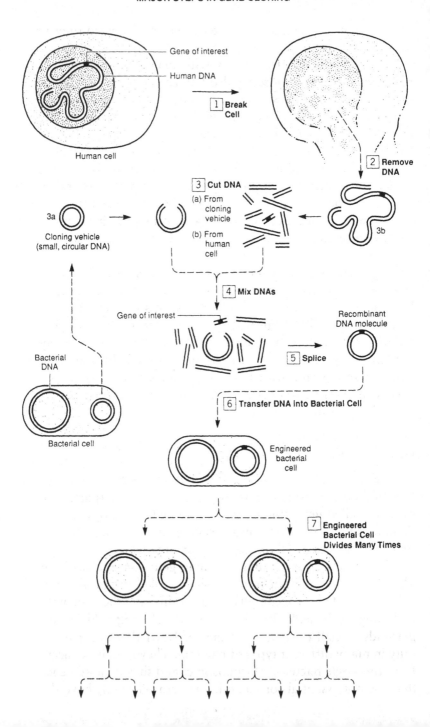

capability to reproduce themselves as the bacterial cell reproduces.

Scientists use restriction enzymes (described in Chapter 3) to splice individual human genes into plasmids. When a circular plasmid is cut open with a restriction enzyme, the ends will bind to fragments of human DNA that have been cut with the same enzyme. A molecule that contains both plasmid DNA and human DNA is called a **recombinant DNA molecule.**

In gene cloning, researchers add plasmids carrying human genes to cultures of bacterial cells. Some of the cells take up the plasmids in a process called **transfection.** Every time a transfected bacterial cell divides, each daughter cell receives one or more copies of the recombinant DNA plasmid. (In the laboratory, many kinds of bacteria divide every 40 minutes. At that rate, and under ideal growing conditions, a single bacterium could generate almost 70 billion new cells in 24 hours. This would provide researchers with billions of copies of the gene of interest.)

The other class of cloning vehicles consists of **bacteriophages** (also called **phages**), which are viruses that infect bacterial cells. Like other viruses, a bacteriophage is a tightly packed core of genetic material (DNA or RNA) wrapped in a protective protein sheath. When a bacteriophage infects a bacterial cell, it takes over the cell's reproductive machinery and directs it to produce new bacteriophage particles. In gene

FIGURE 5.1 Simplified view of the major steps involved in gene cloning. (1) Human cells are broken (for clarity only one cell is shown). (2) DNA containing the gene of interest is removed from human cells. (3) The DNA from a cloning vehicle and human DNA are cut with restriction enzymes. (4) The two types of DNA are mixed. (5) The DNA fragments are spliced together, yielding a recombinant DNA molecule. (6) The recombinant DNA molecule is transferred into a bacterial cell, which has its own DNA. (7) The engineered cell created by step 6 is allowed to reproduce to form a clone of millions of identical cells. SOURCE: Reprinted, with permission, from Karl Drlica, *Understanding DNA and Gene Cloning: A Guide for the Curious* (New York: John Wiley, 1984), figure 1-2. Copyright © 1984 by John Wiley & Sons, Inc.

cloning, researchers use restriction enzymes to replace part of the viral DNA with DNA from another source, for example, a human gene. Bacterial cells infected with these remodeled phages produce copies of the foreign DNA instead of normal viral DNA.

The final step in gene cloning is retrieving the gene copies that have been produced in the bacterial cell. Researchers have several different methods of separating plasmid DNA and bacteriophage DNA from other cell constituents. Once this has been accomplished, restriction enzymes are used to cut the cloned gene copies away from the rest of the plasmid or bacteriophage DNA.

Recent Progress

Researchers have isolated and cloned about 300 human genes. The list includes the normal counterparts of genes responsible for adenosine deaminase deficiency, purine nucleoside phosphorylase deficiency (another severe immune deficiency disease), Lesch-Nyhan syndrome (caused by complete absence of the enzyme hypoxanthine-guanine phosphoribosyltransferase, or HPRT), sickle cell disease, thalassemia, phenylketonuria (PKU), hemophilia A, hemophilia B, alpha$_1$-antitrypsin deficiency, a form of familial hypercholesterolemia, metachromatic leukodystrophy, Gaucher disease, and many others (see Table 5.1).

Articles in the lay press announcing the discovery of a new disease-related gene often refer to the potential for gene therapy resulting from the discovery. Although it is true that identification and cloning of the normal gene are prerequisites for gene therapy, the availability of the gene does not necessarily mean that the associated disease will be an early candidate for gene therapy trials.

Selecting an Appropriate Target Tissue

In the distant future, it may be possible to package cloned genes in such a way that they can be injected directly into the bloodstream or relevant organs of patients who need them. The packaging will protect the foreign DNA from elimination

TABLE 5.1. Examples of single-gene defects for which the mutant gene product is known and the relevant gene has been cloned and mapped.

Disease	Mutant gene product	Chromosomal location of gene
Adenosine deaminase deficiency with severe combined immune deficiency	Adenosine deaminase	20 (long arm)
Alpha₁-antitrypsin deficiency	Alpha₁-antitrypsin	14 (long arm)
Beta thalassemia	Beta chain of hemoglobin	11 (short arm)
Familial hypercholesterolemia (receptor-negative)	Receptor for low density lipoprotein	19 (short arm)
Gaucher disease	Lysosomal glucocerebrosidase	1 (long arm)
Lesch-Nyhan syndrome	Hypoxanthine-guanine phosphoribosyltransferase	X (long arm)
Metachromatic leukodystrophy	Lysosomal arylsulfatase A	22 (long arm)
Ornithine transcarbamylase deficiency	Ornithine transcarbamylase	X (short arm)
Phenylketonuria	Phenylalanine hydroxylase	12 (long arm)
Purine nucleoside phosphorylase deficiency	Purine nucleoside phosphorylase	14 (long arm)
Sickle cell disease	Beta chain of hemoglobin	11 (short arm)
Tay-Sachs disease	Alpha chain of lysosomal hexosaminidase A	15 (long arm)

SOURCE: Human Gene Mapping Library, Howard Hughes Medical Institute, 25 Science Park, New Haven, Connecticut.

by the immune system and direct it to appropriate tissues or cells. For now, however, technical and safety considerations require that transfer of cloned genes into human cells take place outside the body. The first gene therapy studies in human beings will involve the insertion of genes into cells that have been removed from the patient. When these recipient cells are returned to the patient, they will carry the inserted genes with them.

The most basic requirement for a recipient tissue is its

ability to withstand the traumas of removal from the body, manipulation in the laboratory, and reimplantation. In addition, target cells should be capable of dividing and passing their new genetic information on to their descendants. The only two cell types that currently meet these criteria are bone marrow and skin cells.

Bone Marrow Cells

Most gene therapy experiments have focused on bone marrow cells, in part because of the vast experience with bone marrow transplantation from related donors. Scientists have devised standard techniques for withdrawing bone marrow cells from the body and for reintroducing them.

As explained in Chapter 4, bone marrow consists of a heterogeneous mixture of blood cells at different stages of development (see Figure 5.2). Most have a limited lifespan and must be replaced over time. The cell that replenishes the system, the **stem cell,** is the primary target in gene therapy. Stem cells can replicate themselves to maintain the supply of stem cells, or they can differentiate along either of the two major pathways that lead to the formation of more mature cell types. Researchers estimate that one stem cell can generate up to a million mature blood cells.

This tremendous potential for replication is the primary advantage of using stem cells to deliver cloned genes. Theoretically, insertion of a normal gene into a small number of stem cells could result in millions of blood cells producing a normal gene product throughout the body. Moreover, the supply of these cells could last for the lifetime of the patient.

In the early 1980s, optimistic researchers viewed gene therapy as a somewhat more difficult version of existing techniques in bone marrow transplantation. They did not count on the complexity of the blood-forming system.

Only a small proportion—about 1 in 10,000—of bone marrow cells are stem cells. Unfortunately, these cells are difficult to distinguish from many of their more mature descendants. This creates two problems: First, because stem cells are hard to separate from other bone marrow cells prior to

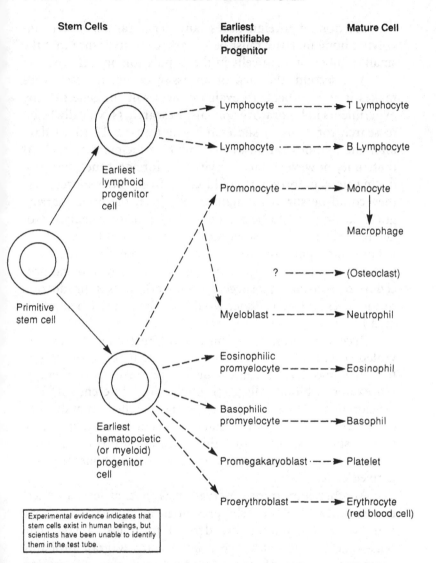

FIGURE 5.2 Blood cell production occurs primarily in the bone marrow. The diagram shows the development of the blood-forming system. Stem cells and early progenitor cells have the capacity for both self-replication and differentiation (production of more specialized progeny). Immediate precursors of the mature blood cells cannot self-replicate—they can only differentiate into mature daughter cells. Broken arrows indicate that there are intermediate steps in development. All mature blood cells except platelets and red blood cells are classified as white blood cells, or leukocytes. (Some researchers believe that microglial cells of the central nervous system also may originate in the bone marrow.)

the insertion of foreign genes, any gene transfer system involving bone marrow must be very efficient to ensure that the small number of stem cells in the population are affected.

And second, the lack of an assay system for stem cells makes it very difficult to evaluate the results of gene therapy experiments in laboratory animals. Researchers typically begin to search for signs of successful gene transfer 12 to 14 days after the administration of marrow to a recipient animal. At that time, however, cells carrying the foreign gene (and producing its gene product) could come from three sources: (1) they could be surviving mature cells from the bone marrow graft; (2) they could be the descendants of intermediate progenitor cells that have some stem cell properties but also have a finite life span; or (3) they could be the descendants of grafted stem cells. Only in the last situation would the procedure be considered a success. (In the other cases, production of the foreign gene product would decline as the mature cells died.)

Recent experiments in mice and larger animals have revealed several other characteristics of the blood-forming system that may have an effect on the future of gene therapy. For example, blood cells go through many different stages of development. At each stage, genes are switched on and off to enable cells to meet new metabolic demands. Evidence from several studies suggests that this maturation process may interfere with the expression of foreign genes by bone marrow-derived cells.

Another important observation is that, at least in mice, the majority of blood cells present in an animal after bone marrow transplantation are derived from just a few stem cells—one to five—although a bone marrow graft probably contains hundreds of such cells. Moreover, the contribution of individual stem cells to blood formation appears to change over time. This finding suggests that the results of gene transfer into bone marrow cells could be quite erratic. If blood formation in a gene therapy patient resulted from the sequential use of a small number of different cell clones (a **clone** is a group of cells derived from a single ancestor), the amount of normal gene product could vary depending on the status of the active clone or clones. If the active clones contained the relevant gene, protein levels might be high enough to overcome the effects of the patient's genetic defect. However, a shift to a clone without the foreign gene would lead to a reduction in the amount of normal protein available, perhaps causing a setback in the patient's condition.

Skin Cells

Two types of skin cells also are being considered as potential target cells for gene therapy: skin fibroblasts and keratinocytes. **Fibroblasts** are the principal cells of connective tissue in the human body. Their main function is to produce collagen and other constituents of the gelatinous mixture that binds cells together. When the skin or another organ is injured, fibroblasts multiply and migrate into the wound, where they deposit new collagen to promote healing. **Keratinocytes** are the primary cells in the protective, outermost layer of the skin, the epidermis. Sheets of cultured keratinocytes have been used extensively to regenerate skin in burn patients.

Gene therapy researchers have had much less experience with skin cells than with bone marrow cells, but several recent studies have produced encouraging results. For example, scientists at the Fred Hutchinson Cancer Research Center and the University of Washington in Seattle successfully introduced an ADA gene into cultured fibroblasts taken from a patient with ADA deficiency. In the test tube, the treated fibroblasts produced 12 times as much ADA enzyme as fibroblasts from normal persons.

The same researchers, in collaboration with scientists at the Salk Institute in La Jolla, California, have begun to explore the use of genetically altered fibroblasts in the treatment of hemophilia B (caused by an inability to produce the blood clotting protein called factor IX). Insertion of factor IX genes into fibroblasts in the laboratory resulted in the production of active factor IX molecules. (Factor IX normally is produced by liver cells.)

The Seattle researchers suggest that fibroblasts might have several advantages over bone marrow cells as delivery vehicles for gene therapy. First, transplanted fibroblasts would not go through the numerous developmental steps required of bone marrow cells, so foreign genes would not be subject to the gene suppression that appears to occur in bone marrow cells. Second, the reintroduction of skin cells, either by injection under the skin or as part of a skin graft, probably would be safer and less expensive than bone marrow transplantation.

The major disadvantage of using fibroblasts for gene transfer is that they have only limited access to the bloodstream (especially compared with the descendants of bone marrow cells). Scientists disagree about whether an enzyme produced solely by fibroblasts could remove a toxic substance from the bloodstream. Fibroblasts might be more effective in the treatment of genetic diseases that result from the absence of a product, such as hemophilia.

For now, this discussion remains speculative because no one has reported trying gene transfer with fibroblasts in an intact animal.[5] It is possible that researchers will discover unanticipated problems with the procedure, similar to those found in experiments involving bone marrow cells. One of many unanswered questions is how long reintroduced fibroblasts will survive in the recipient.

Similar questions surround the use of other types of skin cells in gene therapy, but researchers at the Massachusetts Institute of Technology and Harvard Medical School are enthusiastic about the results of early experiments with keratinocytes. When they inserted a human growth hormone gene into human keratinocytes maintained in laboratory dishes, the treated cells actively secreted growth hormone, although normal keratinocytes do not.

The Boston researchers also transplanted sheets of the treated human keratinocytes onto the backs of laboratory mice (grafts were placed under the mouse skin). The grafts developed into normal-looking epidermal tissue. Analysis of the tissue indicated that the keratinocytes continued to produce low levels of growth hormone after transplantation. However, the researchers were not able to detect human growth hormone in the blood of the experimental animals. Further research is under way to improve the expression of the human growth hormone gene and to study other proteins that might be produced by keratinocytes.

Liver Cells

Several of the more common human genetic disorders, including hemophilia and alpha$_1$-antitrypsin deficiency, affect genes expressed in liver cells. Until recently, liver cells were not considered suitable targets for gene transfer for two reasons: (1) adult liver cells are not susceptible to infection with the types of viruses commonly used in gene transfer experiments, and (2) liver cells could not be removed from and then reintroduced into the body.

Researchers in many laboratories are working to overcome both problems. Dr. Theodore Friedmann at the University of California at San Diego and his coworkers have focused on the insertion of genes into liver cells outside the body. When normally nongrowing liver cells are induced to grow in the laboratory, they undergo a series of developmental changes. For a few days, they revert to a less mature cell type and multiply several times. The UCSD researchers discovered that foreign genes could be inserted into liver cells during this period of growth. The level of foreign proteins produced by treated liver cells is comparable to that produced by other kinds of cells, and the results indicate that inherited liver diseases may be amenable to gene therapy in the future.

Inserting the Gene into Recipient Tissues

Most of the effort in gene therapy research today is devoted to finding the best method for inserting genes into target cells. Ideally, researchers would like to be able to replace a defective gene with its normal counterpart. This would leave the new gene subject to normal regulatory mechanisms—it would be switched on and off at the appropriate times in the cell's life cycle, and it would produce the correct amount of protein each time. However, techniques for inserting DNA into such specific chromosomal sites are still in their infancy and not ready for use in gene therapy.

Today's goals are somewhat more modest: to insert a single copy of the relevant gene into a high proportion of target cells (the site of integration will be different in each cell); to achieve adequate expression of the gene (that is, production of an appropriate amount of protein); and to ensure that the procedure does not harm recipient cells or the patient.

Measures for inserting foreign genes into mammalian cells fall into three categories: techniques using viruses; chemical and physical techniques; and fusion techniques. The consensus among researchers is that modified viruses offer the most promising approach to gene delivery at this time.

Techniques Using Viruses

All viruses can be thought of as packages designed to insert viral genes into host cells. Beyond that basic similarity, however, viruses exhibit as much variation as other organisms. The genome of a virus may consist of RNA or DNA; it may code for only one viral protein or for as many as 50. Each class of viruses has its own way of entering cells and of interacting with the host cell's reproductive machinery.

The viruses considered by most researchers to be best suited to gene therapy are called **retroviruses.** One of the reasons for focusing on retroviruses is that even in their natural state they often carry foreign genes.[6] The basic retrovirus consists of two identical strands of RNA packaged in a **core** of viral proteins (see Figure 5.3). The core is surrounded by a protective coat called the **envelope,** which is derived from the membrane of the previous host cell but modified with glycoproteins (complexes of sugar and protein molecules) contributed by the virus.

Life Cycle of a Retrovirus. The life cycle of a retrovirus begins when it binds to a receptor on the surface of a cell (see Figure 5.4). If a cell does not have a receptor that recognizes particular glycoproteins on the envelope of a retrovirus, the cell is resistant to that virus.

After binding to a receptor, the virus enters the cell and

STRUCTURE OF A RETROVIRUS

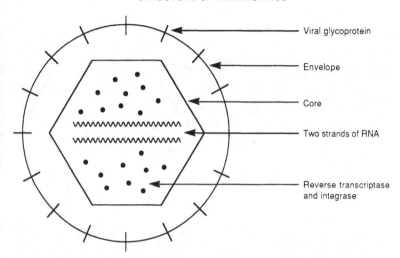

FIGURE 5.3 Schematic diagram of a retrovirus. Two identical strands of RNA are packaged in a core of viral proteins, along with the enzymes reverse transcriptase (which copies the viral RNA into DNA) and integrase (which mediates insertion of the viral DNA into the host cell chromosome). The core is surrounded by the envelope, a protective structure derived in part from the membrane of the previous host cell. The envelope is studded with viral glycoproteins.

sheds its outer envelope. The next step is unique to retroviruses. An enzyme carried inside the viral particle, called **reverse transcriptase,** makes a DNA copy of the viral RNA, using building blocks provided by the host cell. The altered particle then migrates to the nucleus. Another protein, also packaged inside the viral particle, mediates insertion of the new DNA into the host cell chromosome. The integrated viral DNA is called a **provirus.**

The establishment of the provirus is the apex of the retroviral life cycle. Once the provirus is in position, enzymes from the host cell will treat it as an integral part of the cell genome. They will do the work necessary to make new viral particles—copy the viral DNA into RNA and use some of the RNA to produce new viral proteins. Complete progeny viruses

LIFE CYCLE OF A RETROVIRUS

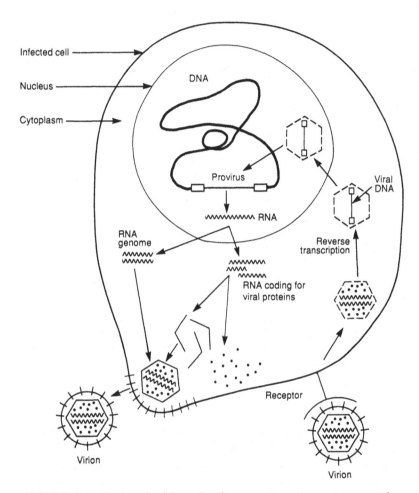

FIGURE 5.4 During the life cycle of a retrovirus, intact virus particles (virions) are taken into the cell via a specific cellular receptor. The enzyme reverse transcriptase, carried inside the viral core, transcribes the single-stranded RNA into double-stranded DNA. The altered particle then migrates to the cell nucleus. Another protein, also packaged inside the viral core, mediates insertion of the new DNA into the host cell chromosome. The integrated viral DNA is called a provirus. In some cases, the provirus remains dormant or unexpressed. In other cases, it is transcribed into viral RNA, leading to the production of viral proteins and the formation of new virus particles. These progeny virions are released by budding from the cell membrane.

are released by budding from the cell membrane (which has been modified with viral glycoproteins).

The most important feature of this life cycle for gene therapy is the retrovirus's highly efficient method of inserting itself into the host cell genome. (In contrast, many other viruses remain separate from the host cell DNA.) Also, infection with a retrovirus appears to result in one provirus per cell. This coincides with the goal of delivering one normal human gene to each target cell in gene therapy. Finally, the presence of the provirus usually does not harm host cells.

Composition of the Retroviral Genome. The small size and relative simplicity of the retrovirus contribute to its value as a delivery vehicle for foreign genes. Figure 5.5 shows the major

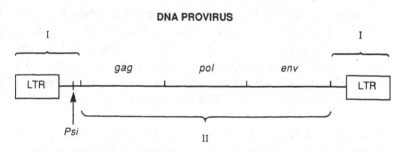

FIGURE 5.5 Simplified structure of the retroviral provirus DNA. The provirus can be divided into two domains. The first domain (I) consists of elements thought to be necessary for transcription and transmission of the viral genome, including the long terminal repeat (LTR) sequences on each end. Adjacent to or within the LTRs are the regulatory signals required for reverse transcription, for insertion of viral DNA into the host cell genome, and for expression of the provirus (transcription of the DNA into messenger RNA and translation of the messenger RNA into protein). A small sequence next to the LTR, labeled *Psi*, is a recognition signal that is required for packaging of the retroviral RNA into an infectious viral particle. The elements of the first domain are retained in the conversion of a wild-type, or natural, retrovirus into a vector for gene transfer. The second domain (II) consists of genes that code for the viral proteins: *gag* codes for the components of the protein core; *pol* codes for reverse transcriptase and for the enzyme required for integration of the provirus into the host cell DNA; and *env* codes for the glycoproteins of the viral envelope. The three viral genes are dispensable from the perspective of a vector because their protein products can be supplied by a helper or packaging cell.

features of a typical retroviral genome in the form of a DNA provirus.

At each end of the provirus are segments called **long terminal repeats** (LTRs). Adjacent to or within the LTRs are the regulatory signals required for reverse transcription, for insertion of viral DNA into the host cell genome, and for **expression** of the provirus (transcription of the DNA into mRNA, and translation of the mRNA into proteins). The sequence within the LTR that controls the transcription of proviral DNA into RNA is called the **promoter.** (It is the initial binding site for the cellular enzyme that copies DNA into RNA.) Another segment of the LTR, the **enhancer,** controls the frequency with which transcription is initiated. A small sequence next to the LTR, labeled *Psi* in the illustration, is a recognition signal that is required for packaging of the retroviral genome into an infectious viral particle.

Between the LTRs are three genes that code for viral proteins. The *gag* gene codes for the components of the protein core; the *pol* gene codes for reverse transcriptase and for the enzyme required for integration of the provirus into the host cell DNA; and the *env* gene codes for the glycoproteins that are incorporated into the outer envelope of the virus.

Preparing a Retroviral Vector.   The effort to turn a natural retrovirus into a delivery vehicle, or **vector,** for human genes is based on several important principles. First, the retroviral vector must have an envelope that is recognized by receptors on the surface of the target cell; otherwise, it will not get into the cell. Second, the vector must have the regulatory signals and the enzymes necessary for copying its genetic information from RNA into DNA, and for inserting the DNA into the host cell genome. Finally, the provirus made by the vector should be structured in such a way that it can be used as a template for the production of the desired human proteins, but not as a template for new viral proteins. The last requirement is very important. If the vector were able to reproduce itself, scientists would not be able to control its activity once target cells had been reimplanted into the body.

The first step in making a suitable vector involves converting a normal provirus (DNA is much easier to work with than RNA) into a recombinant molecule that contains the relevant human gene or genes (see Table 5.2). The most common retrovirus employed for this purpose is the Moloney murine (mouse) leukemia virus, or MoMLV. Using recombinant DNA technology, researchers remove the three retroviral genes from the natural provirus and replace them with the foreign genes (see Figure 5.6).

Along with the human or animal gene of interest, researchers often insert another gene that will allow them to easily identify cells that have incorporated the modified retrovirus. Such a gene is called a **selectable marker**. One that is commonly used is the bacterial gene *neo*, which confers resistance to the antibiotic neomycin. The efficiency of a gene transfer procedure using *neo* as a selectable marker can be determined by exposing recipient cells in a test tube to the drug G418, an analog of neomycin that is highly toxic to mammalian cells. Only cells that are expressing the *neo* gene will survive.

The remodeled provirus, with two LTRs and an intact *Psi* sequence, has the regulatory elements required for inserting its cargo of foreign DNA into the DNA of host cells, but it cannot make the enzymes and other proteins necessary to form

TABLE 5.2. Gene therapy with a retrovirus.

STEP 1. Convert the natural, or wild-type, retroviral provirus into a recombinant DNA molecule—the vector DNA—containing the relevant foreign gene(s).

STEP 2. Make a helper cell to provide functions missing from the vector DNA. (Vector DNA is unable to make enzymes and other viral proteins necessary to form a complete viral particle.)

STEP 3. Introduce the vector DNA into helper cells to produce the viral vector.

STEP 4. Infect target cells (from the patient) with the viral vector to enable cells to produce the protein product(s) of the foreign gene(s). Return cells to the patient.

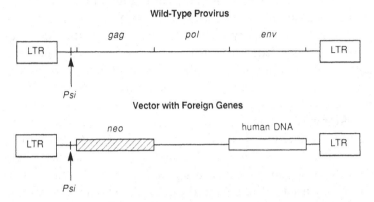

FIGURE 5.6 Scientists use recombinant DNA techniques to replace the *gag*, *pol*, and *env* genes with one or more foreign genes. The foreign genes can be inserted in several different patterns. In this example, a selectable marker gene (*neo*) replaces the viral *gag* and *pol* genes and a human gene replaces the *env* gene.

a complete viral particle. The second step in the development of a gene delivery system addresses this problem. It involves the formation of a **helper cell** or "packaging" cell (see Figure 5.7) that can produce the viral proteins needed to package the recombinant vector.

Within its genome, each helper cell carries an imprisoned "helper" virus. Like the recombinant provirus, the helper provirus is defective; but it has a different type of defect. For example, helper cells developed by researchers at the Whitehead Institute for Biomedical Research in Cambridge, Massachusetts, contain a provirus that is intact except for the *Psi* fragment. The *Psi*-deficient helper virus produces all of the normal viral proteins, but cannot package its own RNA because it lacks the *Psi* recognition signal.

The third step in the gene transfer process, production of the retroviral vector, occurs inside the helper cell (see Figure 5.8). Researchers add copies of the vector provirus to helper cell cultures under conditions that promote the uptake of DNA.[7] In a few cells, the vector provirus integrates into the

STEP 2
MAKING THE HELPER CELL

Designing the Helper Virus

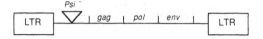

Inserting the Helper Virus into Cells in Culture

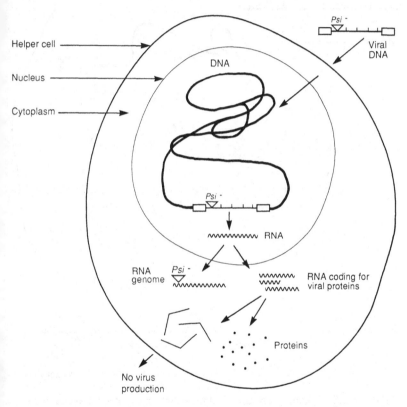

FIGURE 5.7 The helper cell should meet two basic requirements: (1) it should provide functions missing from the vector virus, and (2) it should not be capable of producing viable virus particles. The crucial element in the development of a successful helper cell is the design of the helper virus. Researchers use recombinant DNA techniques to disable the helper virus in the test tube—one of the most common measures is removal of the *Psi* fragment. The *Psi*-deficient helper virus produces all of the normal viral proteins, but cannot package its own RNA because it lacks the appropriate packaging signal. Helper virus DNA is inserted into the genome of the helper cell using chemical techniques.

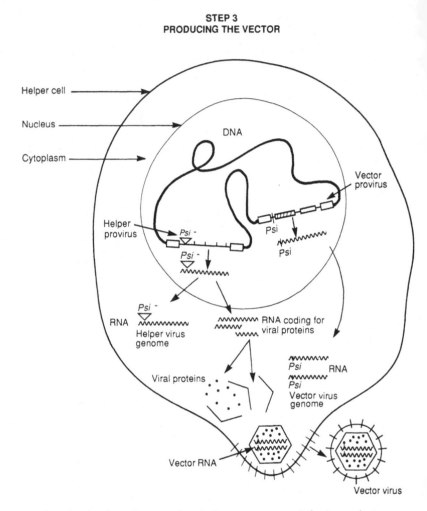

FIGURE 5.8 Scientists use chemical measures or an infection technique to insert recombinant vector DNA (including a human gene) into helper cells. Because the vector provirus contains the *Psi* sequence, the vector RNA genome is automatically encapsulated by viral proteins produced by helper virus DNA in the helper cell. The resulting viral particles are released by budding from the helper cell membrane. The vector virus is capable of only one infection because it lacks the information needed to make viral proteins.

helper cell genome and is copied into RNA. Because the vector provirus contains the *Psi* sequence, the vector RNA is automatically encapsulated by viral proteins always present in the helper cell's cytoplasm (as a result of the presence of the helper virus). The resulting viral particles are then released by budding from the helper cell membrane. Because the helper cell provides the material for the viral envelope, characteristics of the helper cell determine the types of host cells that the retroviral vector can infect.

Viral particles produced in this manner can infect only once, because they lack the information necessary to make viral proteins. An exception would be if copies of helper virus RNA were accidentally encapsulated with the vector RNA. Researchers have devised many techniques to reduce the likelihood of this event, including removing additional elements from the helper virus used to construct the helper cell.

The last step in gene transfer involves exposing appropriate target cells to the vector (Figure 5.9). Researchers either mix target cells with helper cells that are producing the vector virus or bathe them in fluid harvested from the helper cell culture. In either case, the modified retrovirus binds to and enters target cells, undergoes reverse transcription, and integrates itself into the target cell genome. This process of transferring genes via viral vectors is called **transduction.**

Experiments in many laboratories indicate that the concepts underlying the development of retroviral vectors are valid, but numerous questions remain about the best way to design a vector. For reasons that are not completely understood, the insertion of foreign genes often impairs the reproductive capability of retroviruses. The quantity of remodeled virus produced by a helper cell may be much lower than the quantity of natural virus that could be produced by the same cell. Also, elements in the viral LTR may interfere with the expression of human genes carried by the vector. Perhaps the most baffling problems involve interactions between vectors and transduced target cells. For example, transduced bone marrow cells inserted into laboratory animals sometimes appear to shut off the expression of foreign genes over time, but no one knows how or why.

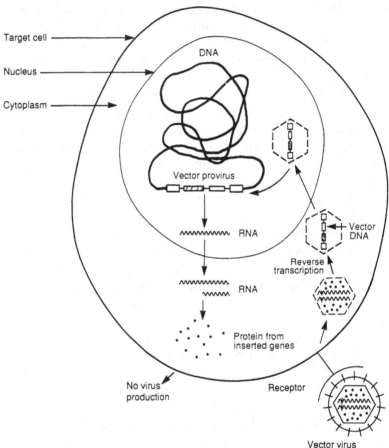

FIGURE 5.9   Researchers infect target cells (such as human bone marrow cells) with the vector virus in two different ways: they mix them with helper cells producing the virus, or they bathe them in fluid harvested from the helper cell culture. When the vector provirus is integrated into the target cell DNA, enzymes from the target cell treat it as an integral part of the cell genome. Cellular enzymes do the work necessary to make proteins from the foreign genes located between the two viral LTRs.

As researchers learn more about the molecular biology of modified retroviruses and about factors that control the expression of mammalian genes, the pace of progress in gene therapy should increase.

DNA Viruses. Gene transfer vectors also have been constructed from several types of DNA viruses (viruses that carry their genetic information in the form of DNA instead of RNA). So far, however, these vectors have not been as successful as the retroviral vectors.

Vectors have been derived from the Epstein-Barr virus (the primary cause of infectious mononucleosis in young adults) and from the bovine (cow) papilloma virus.[8] Both viruses are believed to reproduce in the nucleus in the form of plasmids—they do not integrate themselves into the host cell genome. One theoretical advantage of a system in which the vector remains separate from the host cell's chromosomes would be that the vector would be unlikely to disrupt the function of normal genes in the host cell—either by inactivating a beneficial gene or activating a harmful one. However, scientists do not know whether such a system could result in long-term expression of an inserted gene. The one clear advantage of the Epstein-Barr vector is that it can carry a much larger piece of foreign DNA than a retroviral vector.

William M. Sugden of the McArdle Laboratory for Cancer Research in Madison, Wisconsin, explains that vectors made from herpes simplex viruses and vaccinia viruses—two other types of DNA viruses—are not suitable for gene therapy at this time. Either they kill infected cells or they require the presence of a helper virus inside the target cells (resulting in uncontrolled viral infection).

Early studies of DNA vectors have revealed some additional problems. For example, vectors made from the bovine papilloma virus have a very high mutation rate. In contrast, vectors made from the Epstein-Barr virus are less subject to mutation, but they tend to disappear. Cells lose the Epstein-Barr-derived vectors at the rate of about 4 percent of cells per generation.

DNA vectors have several *theoretical* advantages over

retroviral vectors, but until they can be shown to work efficiently and safely in animal systems, they will not be considered for use in human somatic cell gene therapy. The main stumbling block in working with vectors derived from DNA viruses is lack of information—experience with these vectors is very limited.

Chemical and Physical Techniques

The insertion of foreign genes into the DNA of a mammalian cell is called **transfection** when the DNA enters the cell by chemical or physical means (as opposed to viral means, where the procedure is called transduction). The most common chemical transfection procedure is calcium phosphate-mediated DNA uptake. The two physical techniques employed for gene transfer are microinjection and electroporation.

Calcium Phosphate. In calcium phosphate-mediated DNA uptake, researchers treat target cells with numerous copies of the relevant genes mixed with calcium salts. The foreign DNA travels with the calcium salts, through the cell membrane and cytoplasm into the nucleus. Most cells treated in this way take in the foreign DNA, but very few integrate it into their genomes and begin producing foreign proteins.

Researchers estimate that the transfection rate in bone marrow cells is about 1 in 1,000,000 (the transfection rate in skin fibroblasts is higher—between 1 in 1,000 and 1 in 100,000). The only way to detect the occasional transfected cell is to insert a selectable marker gene (for example, a gene that protects transfected cells from a toxic agent) along with the gene of interest. Even with selection, it is unlikely that procedures using chemical transfection measures can be adapted for use in human gene therapy.

The other drawback of calcium phosphate-mediated DNA uptake is that it often results in the insertion of multiple copies of the transferred gene, all linked together in an unpredictable structure.

Microinjection.   Microinjection is the procedure of choice for many gene transfer experiments in the laboratory because it is highly reliable. Scientists inject DNA directly into the nucleus of the target cell with a very fine glass needle. Up to one cell in five that receives a foreign gene in this way will express it (produce the desired protein).

Despite this advantage, the applicability of microinjection to human somatic cell gene therapy remains questionable. The procedure is extremely laborious—only one cell can be injected at a time. Microinjection studies usually are limited to hundreds or thousands of cells. (In contrast, researchers using viral and chemical techniques can treat billions of cells in one procedure.)

The only situation in which microinjection might be practical for gene therapy is if researchers could distinguish stem cells from other cells in the bone marrow. The transfection of large numbers of stem cells might have an impact on a patient's clinical course.

Electroporation.   Electroporation employs an electrical current to induce target cells to take up DNA and other chemicals from surrounding fluids. Over the past two years, the efficiency of electroporation has increased markedly in several laboratories; however, its potential for human gene therapy remains uncertain.

Fusion Techniques

The final way to get DNA into a target cell is to put it inside a membrane-bound sac, or vesicle. A vesicle can be constructed in such a way that its membrane will fuse with the outer membrane of a target cell. When this occurs in the test tube, foreign DNA from the vesicle spills into the target cell's cytoplasm; some of it may reach the nucleus and integrate into the genome.

Current cell fusion techniques are even less reliable than the chemical techniques described above. In the future, however, researchers hope to be able to construct vesicles that will home in on one particular type of target cell (for example, a liver cell). This might allow direct delivery of foreign genes inside the body.

Conclusions

The basic tasks in gene therapy are isolating the gene responsible for a genetic disease and producing multiple copies of its normal counterpart; selecting target cells in which to deliver the gene to the body; inserting the normal gene into the target cells; and achieving appropriate expression of the inserted gene.

Recombinant DNA technology has made the first task relatively straightforward. More than 300 human genes have been cloned, and the list grows weekly. Researchers have identified the genes responsible for Lesch-Nyhan syndrome, Gaucher disease, chronic granulomatous disease, and many other conditions. They are on the threshhold of isolating the cystic fibrosis gene. But the gap between identifying an abnormal gene and using its normal counterpart to treat a severely ill patient is enormous.

The first clinical trials of gene therapy probably will use a modified retrovirus to insert normal genes into bone marrow cells. Scientists still have many questions, however, about both the target cell and the method of gene insertion. Bone marrow cells go through many different stages of development. At each stage, some genes are activated and others are switched off to allow cells to meet new developmental demands. Recent studies suggest that this maturation process may interfere with the expression of some foreign genes, but the magnitude of the problem remains unclear.

Such experiments underscore the need for more information about the factors that control blood cell development. They also highlight the importance of studying other cell types that might be suitable targets for gene therapy, such as skin cells and liver cells.

Scientists also stress the need for more information about modified retroviruses and other potential mechanisms for inserting genes into cells (such as vectors derived from DNA viruses, microinjection, calcium phosphate-mediated DNA uptake, and fusion techniques). Retroviral vectors have many advantages, but experience with them remains limited. Before current strategies of gene transfer can be employed in human beings, they must be shown to be both safe and effective in laboratory studies. Chapter 6 examines efforts to develop workable animal models of gene therapy.

—— ACKNOWLEDGMENTS ——

Chapter 5 is based on the presentations of W. French Anderson, Mario R. Capecchi, Theodore Friedmann, David W. Martin, Jr., A. Dusty Miller, Richard Mulligan, Stuart Orkin, William M. Sugden, and Harold E. Varmus.

—— NOTES ——

1. A rat gene for growth hormone was injected into fertilized eggs from a mouse strain deficient in growth hormone. Animals originally destined to be dwarfs grew to almost twice the size of a normal mouse. For reasons explained in Chapter 1, this approach (germ line gene therapy) is not considered appropriate for use in human beings.

2. Anderson, W. French. "Beating Nature's Odds." *Science 85,* November 1985, pp. 49-50.

3. Normal persons have two copies of the retinoblastoma gene. The protein product of the gene appears to protect cells in the eye against unregulated growth. Children with a hereditary susceptibility to retinoblastoma have just one copy, and inactivation of the single copy by environmental or other factors leads to eye cancer.

4. Researchers also have developed cloning techniques that employ mammalian cells in tissue culture; these are used primarily to produce proteins that would not be manufactured correctly by bacteria or yeast cells.

5. In 1987, Richard Selden and his coworkers at Massachusetts General Hospital in Boston described a model for somatic cell gene therapy in which cultured mouse fibroblasts were transfected with a human growth hormone (hGH) gene and cells from one of the resulting clonal lines were subsequently implanted into various locations in mice. This approach, called transkaryotic implantation, represents a possible alternative to gene transfer with retroviruses.

The method's potential strengths and weaknesses are described in *Science* 236:714-718.

6. Researchers studying retroviruses that infect chickens, rodents, cats, and monkeys have found that many of them carry genes originally derived from animal cells. Such genes, called proto-oncogenes, can, under certain circumstances, trigger cancerous growth in infected cells. The retroviruses used for gene therapy do not carry proto-oncogenes. Chapter 6 explores this and other safety issues in more detail.

7. The retroviral vector also can be inserted into helper cells using an infection technique. See, for example, Miller et al., *Somatic Cell and Molecular Genetics* 12:175-183.

8. Some human counterparts of the bovine papilloma virus cause common warts, and some are associated with cervical cancer.

—— SUGGESTED READINGS ——

Anderson, W. French. 1984. "Prospects for Human Gene Therapy." *Science* 226:401-409.

——— 1985. "Beating Nature's Odds. *Science 85*, November, pp. 49-50.

Dick, John E., Maria Cristina Magli, Robert A. Phillips, and Alan Bernstein. 1986. "Genetic Manipulation of Hematopoietic Stem Cells with Retrovirus Vectors." *Trends In Genetics* 2:165-170.

Drlica, Karl. 1984. *Understanding DNA and Gene Cloning: A Guide for the Curious*. New York: John Wiley.

Joyner, Alexandra, Gordon Keller, Robert A. Phillips, and Alan Bernstein. 1983. "Retrovirus Transfer of a Bacterial Gene into Mouse Haematopoietic Progenitor Cells." *Nature* 305:556-558.

Kolata, Gina. 1985. "Closing In on the Muscular Dystrophy Gene." *Science* 230:307-308.

——— 1986. "Two Disease-Causing Genes Found." *Science* 234:669-670.

Mann, Richard, Richard C. Mulligan, and David Baltimore. 1983. "Construction of a Retrovirus Packaging Mutant and Its Use to Produce Helper-Free Defective Retrovirus." *Cell* 33:153-159.

Marx, Jean L. 1987. "Probing Gene Action during Development: The Good News—and the Bad—about Gene Therapy Prospects." *Science* 236:29-30.

McCormick, Douglas. 1985. "Human Gene Therapy: The First Round." *Bio/Technology* 3(8):689-693.

Miller, A. Dusty, Robert J. Eckner, Douglas J. Jolly, Theodore Fried-mann, and Inder M. Verma. 1984. "Expression of a Retrovirus Encoding Human HPRT in Mice." *Science* 225:630-632.

Miller, A. Dusty, David R. Trauber, and Carol Buttimore. 1986. "Factors Involved in Production of Helper Virus-Free Retrovirus Vectors." *Somatic Cell and Molecular Genetics* 12(2):175-183.

Morgan, Jeffrey R., Yann Barrandon, Howard Green, and Richard C. Mulligan. 1987. "Expression of an Exogenous Growth Hormone Gene by Transplantable Human Epidermal Cells." *Science* 237:1476-1479.

Orkin, Stuart H. 1986. "Reverse Genetics and Human Disease." *Cell* 47:845-850.

Palmer, Theo D., Randy A. Hock, William R. A. Osborne, and A. Dusty Miller. 1987. "Efficient Retrovirus-Mediated Transfer and Expression of a Human Adenosine Deaminase Gene in Diploid Skin Fibroblasts from an Adenosine Deaminase-Deficient Human." *Proceedings of the National Academy of Sciences* 84:1055-1059.

Selden, Richard F., Marek J. Skoskiewicz, Kathleen Burke Howie, Paul S. Russell, and Howard M. Goodman. 1987. "Implantation of Genetically Engineered Fibroblasts into Mice: Implications for Gene Therapy." *Science* 236:714-718.

Williams, David A., Ihor R. Lemischka, David G. Nathan, and Richard C. Mulligan. 1984. "Introduction of New Genetic Material into Pluripotent Haematopoietic Stem Cells of the Mouse." *Nature* 310:476-480.

6 | Progress Report: Experiments in Animals

When will it be ethical to begin gene therapy in human beings? W. French Anderson and John C. Fletcher of the National Institutes of Health addressed this issue in a 1980 article in the *New England Journal of Medicine*.[1] They concluded that before beginning gene therapy trials in human beings, researchers must show in animal studies that (1) the new gene can be put into the correct target cells and will remain there long enough to be effective; (2) the new gene will be expressed in the cells at an appropriate level; and (3) the new gene will not harm the cell or, by extension, the animal.

These requirements are very similar to those applied to other new forms of therapy. Basically, they ask researchers to demonstrate through animal models that the probable benefits of therapy will outweigh the probable risks.

Animal Models

Animal models of human diseases allow researchers to design and evaluate new therapeutic strategies that could not be assessed adequately in clinical trials because of the limitations on human experimentation. Two different types of animal models are used in the study of inherited diseases. The first is an animal that has a naturally occurring disease analogous to a recognized human disease. The second is an animal in which an abnormal condition is artificially induced to permit tests of a potential therapeutic measure.

Animal Analogues of Human Inherited Metabolic Diseases

Over the past 15 years, researchers have identified hundreds of naturally occurring models of inherited human diseases in mice and domestic animals (see Table 6.1). More than 30 models exist of human lysosomal storage disorders, including cats with Tay-Sachs disease, Hurler syndrome, and Maroteaux-Lamy syndrome (see Figure 6.1). Scientists also are studying dogs with hemophilia A and B; rabbits with a defect resembling familial hypercholesterolemia; and mice with beta thalassemia.[2]

TABLE 6.1.   Natural animal models of some inherited human diseases.

Disorder	Animal
Carbohydrate metabolism	
Diabetes mellitus	Cat, dog, guinea pig, hamster, monkey, mouse, rat
Galactosemia	Rat
Mannosidosis	Cat, cattle, goat
Pituitary dwarfism	Dog, mouse
Amino acid and organic acid metabolism	
Albinism	Chicken, frog, hamster, mouse
Chediak-Higashi syndrome	Cat, cattle, mink, mouse
Citrullinemia	Dog
Ornithine transcarbamylase deficiency (sparse fur)	Mouse
Lipid metabolism	
G_{M2} gangliosidosis (Tay-Sachs disease or Sandhoff disease)	Cat, dog, pig
Gaucher disease	Dog
Hypercholesterolemia (affecting LDL receptor)	Rabbit
Metachromatic leukodystrophy	Mink
Niemann-Pick disease	Cat, dog, mouse

TABLE 6.1.   (Continued)

Disorder	Animal
Mucopolysaccharide metabolism	
Hurler syndrome	Cat, dog
Maroteaux-Lamy syndrome	Cat
Porphyrin and heme metabolism	
Crigler-Najjar syndrome	Gunn rat
Erythropoietic porphyria	Cat, cattle, pig, squirrel
Connective tissue, muscle, and bone	
Ehlers-Danlos syndrome	Cat, cattle, dog, mink, mouse, sheep
Osteopetrosis	Cattle, mouse, rabbit, rat
Blood and blood-forming tissues	
Alpha thalassemia	Mouse
Beta thalassemia	Mouse
Transport	
Cystinuria	Dog, mink
Fanconi syndrome	Dog
Circulating enzymes and plasma proteins	
Hemophilia A	Cat, dog
Hemophilia B	Cat, dog
von Willibrand disease	Dog, pig, rabbit

SOURCE: Adapted, with permission, from George Migaki, "Compendium of Inherited Metabolic Diseases in Animals," in Robert J. Desnick, Donald F. Patterson, and Dante G. Scarpelli, eds., *Animal Models of Inherited Metabolic Diseases* (New York: Alan R. Liss, 1982).

See also *Animal Models of Human Disease* (looseleaf, 1972-present). Registry of Comparative Pathology, Armed Forces Institute of Pathology, Washington, D.C. 20306.

An important part of assessing each new animal model is to determine how closely the animal disease is related to the human disease. Diseases that produce identical physical symptoms in two species may be very different at the biochemical level.

FIGURE 6.1 (Left) A normal Siamese cat at age 3 years—note the clear corneas and the shape of the nose. (Below) Littermates with the inherited disease Maroteaux-Lamy syndrome at age 6 months—note the corneal clouding, the depressed nasal bridge, and the large forepaws. SOURCE: Reprinted, with permission, from Robert J. Desnick et al., "Enzyme Manipulation Therapy: A Novel Strategy for the Treatment of Inborn Errors of Metabolism," in Raul A. Wapnir, ed., *Congenital Metabolic Diseases: Diagnosis and Treatment* (New York: Marcel Dekker, 1985), p. 269.

The crucial question for inborn errors of metabolism is whether the animal disease and the human disease are caused by the same enzyme or other protein. If different proteins are involved, the animal disease may not be an appropriate model for understanding the human disease or for evaluating new therapeutic measures.

Increasingly, new techniques in genetic engineering will enable scientists to design animal models of human genetic diseases that do not have a natural animal counterpart. Michael Kuehn and his colleagues at the University of Cambridge recently developed mutant mice that have the same biochemical defect as patients with Lesch-Nyhan syndrome.[3] They used multiple infections with a retroviral vector to increase the frequency of mutations in cultured mouse embryonic cells, and then selected for cells that had a mutation in the *HPRT* gene. The selected cells were implanted in mouse embryos, where they contributed to the development of chimeric mice (mice containing cells with two different genetic origins). Some of the offspring of the chimeric mice inherited the mutant gene; males with the mutant gene had no HPRT activity.

Early studies indicate that the mutant mice do not develop the neurological and behavioral symptoms common to patients with Lesch-Nyhan syndrome. Some researchers suspect that the mice may have an alternative metabolic pathway that compensates for the absence of HPRT. Nonetheless, the successful design of HPRT-negative mice represents an important step in the field of molecular genetics.[4]

Animal analogues of human genetic diseases provide invaluable information about important biochemical processes and about the potential value of certain therapeutic measures, but they have limitations. Species differences at the level of the gene defect and at the level of cell function may affect the way in which animals and human beings with similar inherited disorders respond to a given therapeutic strategy.[5] Although studies in animal models are an important prelude to human trials in genetic disease research, they will never replace human studies.

Animal Models in Somatic Cell Gene Therapy

The criteria set forth by Drs. Anderson and Fletcher do not require researchers to demonstrate that gene therapy can cure a genetic disease in animals. Instead, the criteria focus on the *process* of gene therapy—the new gene must remain in target cells "long enough to be effective" and must be expressed in target cells "at an appropriate level."

This emphasis on process is important, because animal analogues have not been identified for most of the diseases likely to respond to gene therapy in humans. For example, researchers do not have a natural animal model for adenosine deaminase (ADA) deficiency. They cannot insert a normal *ADA* gene into an ADA-deficient animal and assess its effects, because no such animal is known.

The Basic Protocol. To overcome this problem, scientists have developed an alternative experimental strategy using normal animals. The protocol varies somewhat depending on the species involved and the nature of the experiment, but it usually begins with the removal of bone marrow from the animal (see Figure 6.2).[6] The marrow is infected in the laboratory with a retroviral vector carrying copies of the gene of interest. (This process is described in detail in Chapter 5.) Meanwhile, the animal is treated with enough radiation to destroy all of its remaining bone marrow. After the radiation treatment, the infected marrow cells are put back into the animal, where they multiply and repopulate the blood-forming system. Several different methods are used to determine how many of the new blood cells carry the transferred gene and whether they are expressing it (using it to produce protein).

Destruction of the animal's remaining marrow is necessary in this system because the infected cells usually do not have any competitive advantage over other bone marrow cells in the animal's body. For example, ADA studies often involve the insertion of a human *ADA* gene into animal cells that carry copies of the normal animal *ADA* gene. The ability to produce human ADA does not increase the survival potential of these cells in any way. If the infected cells were injected back into an intact animal, they would disappear over time. Elimination of the animal's existing bone marrow by radiation provides space for the treated cells to grow. (However, if the number of bone marrow cells returned to the animal is too low, the animal will die.)

GENE TRANSFER PROTOCOL IN THE MOUSE

Remove bone marrow
from donor and isolate
marrow cells

Donor mouse

Infect marrow cells
with retroviral vector

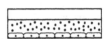

Helper cell culture
producing
retroviral vector

Fluid harvested
from helper
cell culture

Infuse treated marrow
cells into recipient

Recipient mouse .
(radiation used to destroy
blood-forming system before
infusion of donor cells)

Short-Term Assay

Twelve to fourteen days
after the transplant,
scientists examine spleen
cells for evidence of the
inserted gene(s).

Long-Term Assay

Four months or more after
the transplant, scientists
examine blood cells from
many tissues in the body
for evidence of the inserted
gene(s).

Mice or Larger Animals? The animal used most often in gene transfer studies is the mouse. The mouse model has several advantages over other animal systems. First, scientists have more experience with bone marrow transplantation in the mouse than they have with transplantation in other animals. In addition, the blood-forming system in the mouse has been well characterized; researchers can identify all but the most primitive mouse stem cells. Finally, the mouse is easy to manage. Mice are hardy and reproduce rapidly. Researchers can collect statistically significant data in a relatively short period of time.

Some scientists believe that the limited resources available for gene transfer studies should be spent exclusively on mouse research until more data have been collected, but others believe it is important to begin assessing more complex models. Researchers have now conducted preliminary gene transfer experiments in dogs, sheep, and monkeys. One of the questions addressed by these experiments is whether protocols developed in mice can be scaled up sufficiently to meet the needs of larger animals (and, eventually, human beings). About one million bone marrow cells are required to repopulate the bone marrow of a mouse; in dogs and monkeys, the figure is close to one billion.

FIGURE 6.2 Gene transfer into bone marrow cells in the mouse model. Bone marrow cells are removed from a normal animal or from an animal treated with drugs known to increase the proportion of stem cells in the body. A retroviral vector (described in Chapter 5) is used to insert foreign genes into the bone marrow cells. The cells are infected with the vector in one of two ways: they are mixed with helper cells producing the vector, or they are bathed in fluid harvested from the helper cell culture. After infection, the treated cells are infused into a mouse whose blood-forming system has been destroyed by radiation. (In mice, the donor and the recipient are usually different, but genetically identical, animals; in large animals, the treated cells would be injected back into the donor.) Long-term survival of the recipient animal depends on the introduction of enough bone marrow cells to repopulate the blood-forming system.

Animal Models in Germ Line Gene Therapy

The animal experiments that have received the most attention in the lay press have been those involving germ line gene therapy. Using microinjection techniques, researchers have inserted functional genes into embryos from several different mouse strains. In September 1986, scientists at Columbia University described their successful efforts to treat mouse beta thalassemia by transferring cloned beta-globin genes into mouse embryos lacking a beta-globin gene. More recently, scientists at the California Institute of Technology and Harvard Medical School used germ line gene therapy to correct a mouse neurological disorder that causes uncontrollable shivering, convulsions, and early death. (This disorder does not have a recognized human counterpart.)

These experiments demonstrate that placing a functional gene inside an intact animal can cure a genetic disease, but their relevance to human somatic cell gene therapy is limited. For reasons described in Chapter 1, germ line gene therapy is not considered appropriate for use in humans. Moreover, fertilized eggs and early embryos respond differently to gene transfer than cells at later stages of development.

Inserting Genes into Somatic Cells

Over the past three years, researchers have used retroviral vectors to insert genes into many different types of cells in culture: bone marrow cells from mice, dogs, and human beings; lymphoid cells from patients with HPRT deficiency (Lesch-Nyhan syndrome) and ADA deficiency; liver cells from rats; and human skin fibroblasts and keratinocytes. The first successful experiments created high expectations. Many in the field believed that the experiments would lead to a set of rules—rules defining the ideal vector for inserting specific genes into human beings.

The search for these rules continues, but expectations are considerably more restrained than they were in 1985. Most researchers no longer expect to find a single vector that works well in every situation. The variables that influence a vector's success or failure are just beginning to be understood. For example, the gene inserted into a vector has a much bigger influence on the vector's activity than scientists originally expected. In several early studies involving mouse cells, vectors produced very good results when used to transfer the *neo* gene (which confers resistance to drugs of the neomycin family), but produced very poor results with the *ADA* gene.

The cell's environment also appears to play an important role in the outcome of gene transfer experiments. Researchers in many laboratories have found that vectors that express gene products efficiently in cell culture do not do so in live animals. For example, A. Dusty Miller and his colleagues at the Fred Hutchinson Cancer Research Center successfully transferred both the *neo* gene and the *DHFR* gene (which codes for dihydrofolate reductase and confers resistance to the highly toxic drug methotrexate) into canine bone marrow cells in the laboratory.[7] Proteins produced by the foreign genes enabled the cells to grow in culture media containing the toxic drugs. But when the researchers tried to apply their cell culture experience to bone marrow transplantation in dogs, the results were disappointing. Although the dogs' blood-forming systems were reconstituted by the transplanted marrow (showing that the procedure could be scaled up appropriately), the scientists found only fleeting evidence of the expression of the drug resistance genes. (That is, they did not detect any vector DNA sequences in cells from the animals.)

Long-term studies of mice have produced more encouraging results. In late 1985 and early 1986, three groups of researchers in Canada, Europe, and the United States reported persistence of the *neo* gene in mice that had received bone marrow transplants. These studies are significant for several reasons. First, they demonstrate that the general protocol envisioned for eventual gene therapy experiments in human beings is feasible. Second, they show that genes can be inserted into mouse stem cells and that the cells will function normally.[8]

The importance of the long-term assay is underscored by Richard C. Mulligan, one of the pioneers in the field of gene transfer with retroviral vectors. Unlike conventional short-term assays, the long-term assay can distinguish between stem cells and more mature bone marrow progenitor cells that have some stem cell properties but also have a finite life span. Successful gene therapy will require insertion of the relevant gene into the most primitive stem cells. Thus, long-term studies, in which mice are examined four to five months after transplantation, provide a more accurate picture of the potential for gene therapy in human beings.[9]

Expression of Inserted Genes

The ability to infect a large proportion of target cells is one measure of the efficiency of a gene transfer procedure. The other is how well the gene functions once it has been inserted into a target cell. Gene expression depends on the structure of the retroviral vector and on its environment within the cell.

Designing Vectors to Improve Expression

Retroviral vectors used in the first gene transfer studies were primarily **double expression vectors.** An example of this type of vector is shown in Figure 6.3. One foreign gene (usually a selectable marker gene) replaces the viral *gag* and *pol* genes, and another replaces the viral *env* gene. The distinguishing feature of these vectors is that the regulatory elements in the viral long-terminal repeat (LTR) play a dual role. In the helper cell,[10] they switch on transcription of the entire retroviral provirus to generate RNA copies needed for the formation of infectious virus particles. In the target cell, they regulate the expression of proteins from the two foreign genes.

Measurements of gene expression in cell culture studies and in short- and long-term animal studies indicate that dou-

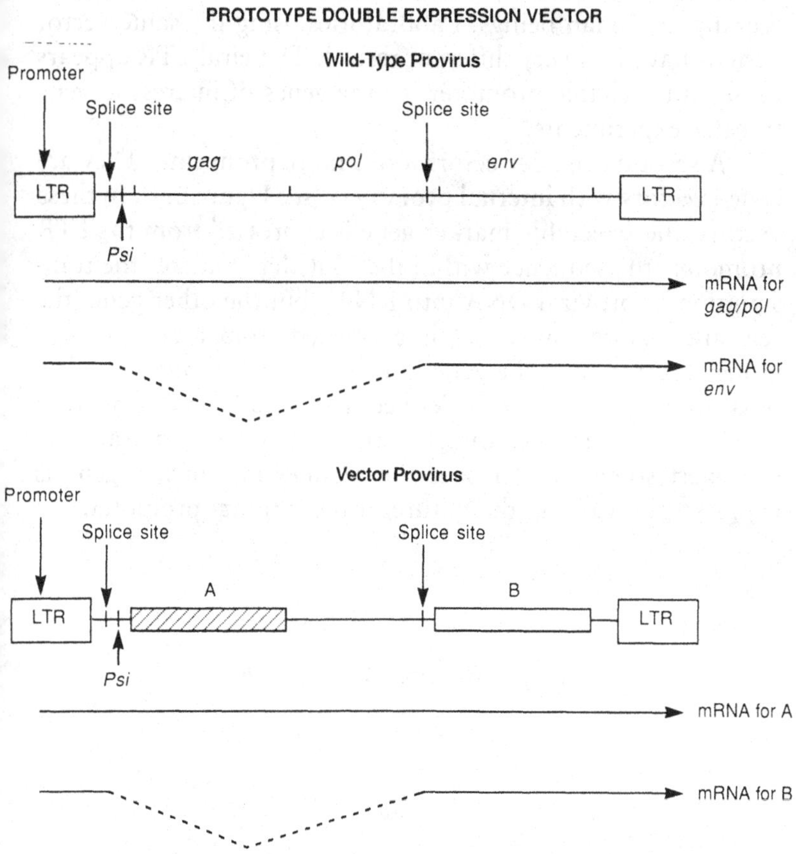

FIGURE 6.3. The design of the double expression vector is based on the unique features of retroviral RNA expression. The retrovirus contains one promoter, but generates two types of messenger RNA (mRNA) that accumulate in the cytoplasm of the infected cell. The *gag* and *pol* genes are expressed from the full-length mRNA molecule, and the *env* gene is expressed from a shorter, spliced mRNA molecule (splicing is performed by enzymes made by the host cell). When the retrovirus is used to make a vector provirus, one foreign gene (**A**) replaces the *gag* and *pol* genes, and the other (**B**) replaces the *env* gene. The first foreign gene is expressed from the unspliced mRNA, and the second is expressed from the spliced mRNA. SOURCE: From Eli Gilboa, "Retrovirus Vectors and Their Uses in Molecular Biology," *BioEssays* 5:252-257. Copyright © 1987 by Cambridge University Press. Reprinted with permission.

ble expression vectors probably will not be suitable for gene
therapy in human beings. Laboratories using the same vector
designs have had very different results. The viral LTR appears
to be an unreliable promoter for the genes of interest in gene
transfer experiments.

A second class of vectors looks more promising. They are
called **vectors with internal promoters** (see Figure 6.4). In these
vectors, the selectable marker gene is expressed from the LTR
promoter (the sequence within the LTR that controls the transcription of proviral DNA into RNA), but the other gene (the
relevant mammalian gene) is expressed from a special promoter inserted with the gene into the retroviral provirus. The
separate promoter may be derived from another virus or from
a cellular gene. The advantage of this system is that researchers
can exert some control over the manner in which a gene is
expressed by varying the nature of the internal promoter.

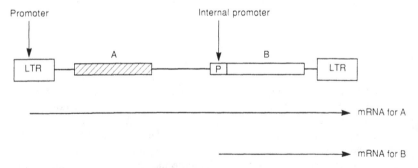

VECTOR WITH INTERNAL PROMOTER

FIGURE 6.4   In the vector with internal promoter, one foreign gene (**A**)—
usually a selectable marker gene—is expressed from the LTR promoter. The
other foreign gene (**B**) is expressed from a separate promoter linked to the
gene before insertion into the retroviral provirus. The internal promoter
may be derived from another virus or from a cellular gene. SOURCE: From
Eli Gilboa, "Retrovirus Vectors and Their Uses in Molecular Biology,"
BioEssays 5:252-257. Copyright © 1987 by Cambridge University Press.
Reprinted by permission.

A retroviral vector called N2, developed by Eli Gilboa of Memorial Sloan-Kettering Cancer Center and his colleagues, has been used successfully to construct vectors with internal promoters in several laboratories. Randy Hock and A. Dusty Miller of the Fred Hutchinson Cancer Research Center employed N2 to transfer the *neo* and *DHFR* genes into human and canine bone marrow cells. Philip W. Kantoff of the National Institutes of Health and his coworkers inserted an *ADA* gene (linked to a promoter from the primate virus SV40) into N2 and used it to correct ADA deficiency in cultured T lymphocytes from patients with severe combined immune deficiency. In addition, two research groups have used vectors generated from N2 to achieve efficient expression of the *neo* gene in the mature blood cells of "fully reconstituted" mice (whose continued survival is dependent on the descendants of stem cells from a bone marrow graft).

These results notwithstanding, Gilboa cautions that N2 is not the highly efficient, all purpose vector that gene therapy scientists set out to find four years ago. He notes that researchers have a tendency to publish only positive results, and that this tendency can lead to misconceptions about the potential benefits of a new technique. Not all genes inserted into the N2 vector have been properly expressed; efforts are under way to determine why some genes work better than others in the N2 construct.

One potential drawback of all vectors with internal promoters is that the LTR promoter and the internal promoter may work against each other. If the internal promoter inhibits the activity of the LTR promoter, helper cells may not produce enough virus particles to adequately infect target cells. On the other hand, if the LTR promoter suppresses the activity of the internal promoter, the expression of the inserted gene may be too low to be clinically useful.

A third type of retroviral vector, called a **self-inactivating vector,** may overcome at least one of these problems. The unique feature of the self-inactivating vector is that the promoter and another regulatory element, the enhancer (the segment of the provirus that controls the frequency with which transcription is initiated), "disappear" from both viral LTRs during the integration of the provirus into the target cell chromosome (see Figure 6.5).[11] The LTR promoter is available to promote replication of the vector in the helper cell, but it is not present to interfere with gene expression in the target cell. Research on the properties of self-inactivating vectors has just begun, but scientists have shown that the self-inactivation mechanism works.

SELF-INACTIVATING VECTOR

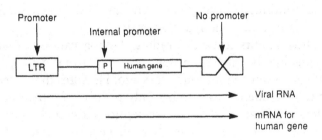

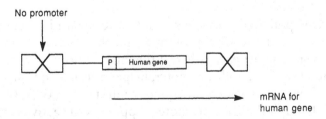

FIGURE 6.5 Production of the self-inactivating vector begins with vector DNA that has an inserted gene linked to an internal promoter, one intact LTR, and one LTR from which the promoter and enhancer sequences have been deleted. When the viral RNA is copied into DNA in the target cell (see the description of the life cycle of a retrovirus in Chapter 5), the deletion in the right LTR is reproduced in the left LTR. As a result, the only active promoter in the new provirus is the internal promoter. SOURCE: From Eli Gilboa, "Retrovirus Vectors and Their Uses in Molecular Biology," *BioEssays* 5:252-257. Copyright © 1987 by Cambridge University Press. Reprinted with permission.

One of the reasons for focusing on self-inactivating vectors is that they may be the most effective means of transferring whole genes into cells. Most of the "genes" inserted into viral vectors actually have been pieces of cDNA that code for the desired gene product. There are several differences between these cDNA fragments (produced from messenger RNA, as described in Chapter 5) and the corresponding chromosomal genes. The most obvious is that the whole gene possesses its own promoter. In addition, the whole gene may have one or more noncoding regions called **introns.**

In the late 1970s, researchers discovered that the coding sequences of many genes in higher animals are interrupted by DNA sequences that do not code for protein subunits (see Figure 6.6). After the gene is transcribed into RNA, these noncoding regions, or introns, are cut out and the coding regions are rejoined. Such "spliced" RNA molecules become the messenger RNAs that are translated into protein. (Because messenger RNA is used as a template to synthesize cDNA, cDNA molecules also lack introns; see Figure 6.7) Initially, the role of the intron was a mystery, but it has now been determined that some introns contain regulatory elements that work in conjunction with the promoter to ensure appropriate expression of the gene's coding segments.

The ultimate objective of gene therapy is to provide target cells with genes that behave just as they would in their natural chromosomal contexts. Scientists believe that vectors carrying and expressing whole genes (often called **genomic vectors**) will be more likely to accomplish this task than existing vectors.[12] But the development of genomic vectors is very complex. All of the regulatory elements present in the whole gene have the potential to interact with regulatory elements in the viral LTRs. The advantage of using self-inactivating vectors to carry whole genes is that the deletions in the LTRs reduce possible interactions between the retroviral genome and the genome of the inserted gene.

Measuring Expression

In cell culture studies, expression usually is determined by the ability of cells to survive under conditions that require pro-

EXONS AND INTRONS

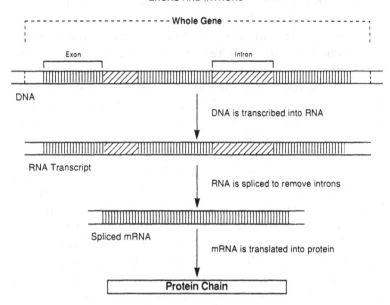

FIGURE 6.6 Many genes in higher animals and plants consist of bits of coding sequences, called exons (or expressed regions), separated from one another by noncoding sequences called introns (or intervening sequences). Transcription of such a gene into RNA proceeds through both exons and introns, but only the exons contain information pertaining to the structure of the gene's protein product (the order of amino acids in the protein chain). Thus, the introns must be removed before the RNA can be translated into protein. Specific enzymes in the cell nucleus cut out the introns and rejoin the exons. The resulting "spliced" RNA molecule (now a coherent piece of messenger RNA) is transported from the nucleus to the cytoplasm, where it is translated into protein. SOURCE: Adapted, with permission, from Stanbury et al., *The Metabolic Basis of Inherited Disease*, 5th ed. (New York: McGraw-Hill, 1983), p. 1 and figure 2-7.

duction of the protein product. As we have seen, when mammalian cells that have been infected with a vector containing the *neo* gene are put in a medium containing the toxic drug G418, only cells that produce the *neo* gene protein will survive. The problem with this assay is that it provides no information regarding the actual level of gene expression. The level of *neo* gene product required to make a cell resistant to G418 is extremely low. A vector may generate enough protein from

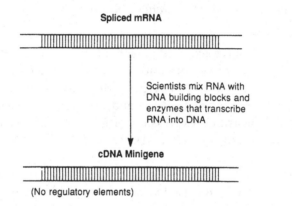

FIGURE 6.7   The whole gene possesses its own promoter and one or more introns. The introns may contain regulatory elements that work in conjunction with the promoter to ensure appropriate expression of the gene's coding segments. In contrast, the cDNA minigene (produced from messenger RNA) contains only the coding segments (or exons) of the gene.

the *neo* gene to permit cell survival, but when the same vector construct is used to carry a mammalian gene, the amount of protein that gene produces may turn out to be insignificant.

Assays used in studies of intact animals provide more complete information. They involve direct measurements of the relevant gene product or of messenger RNA coding for that product. The gene product is extracted from the blood cells of animals that have received bone marrow transplants according to the procedures described above. These tests allow researchers to determine how the level of gene expression obtained with experimental vectors compares with the level of expression that would be required to correct a genetic defect in human beings.

Most data on gene expression have come from short-term studies of transplant recipients, but the best indicator of a vector's potential for human therapy is how it performs over the long term. Only long-term studies (at least four months in mice) provide conclusive evidence that a vector functions efficiently in the descendants of the most primitive stem cells.

Expression of Inserted Genes—Results

Genes introduced into cultured cells often work very well. Researchers have achieved reasonable levels of expression with both LTR-based vectors and vectors with internal promoters in a wide range of cell types. Most studies have involved selectable marker genes (*neo* and *DHFR*), but scientists also have demonstrated expression of genes more relevant to human gene therapy, such as *ADA* and *HPRT*.

Mouse Studies. The results in mouse studies have been quite varied. A few LTR-based vectors have yielded reasonable levels of expression in short-term studies of transplanted mice, but the results have not been consistent. Researchers conducting long-term studies of LTR-based vectors have reported dramatic declines in the expression of the *neo* gene over time.[13]

The most encouraging data have come from experiments involving vectors with internal promoters. Researchers have tested promoters from several different viruses and from human genes. The herpes virus thymidine kinase promoter (the DNA sequence that initiates expression of the thymidine kinase gene) has been most effective in promoting expression of the *neo* gene in the mouse model.

Unfortunately, only a few reports have been published describing gene expression in fully reconstituted mice. Accurate assessments of future retroviral vectors will require increased emphasis on experiments that follow treated animals for long periods.

Only one long-term study has addressed the issue of

whether whole genes inserted into bone marrow cells will function in appropriate somatic cells in the body. Researchers at the Whitehead Institute and MIT transferred the human beta-globin gene into mouse bone marrow cells and introduced the cells into mice via bone marrow transplantation. Studies of three recipients engrafted for more than four months were very encouraging. The investigators found significiant levels of human beta-globin RNA in the blood of all three mice tested. Expression of the gene was low compared with that of the mouse beta-globin gene (5 to 10 percent), but the construct appeared stable. The most exciting finding was that expression of the beta-globin gene in one mouse was restricted to blood, bone marrow, and spleen cells. No beta-globin RNA was detected in other cell types descended from the transduced stem cells. This is promising, because the application of gene therapy to human hemoglobin disorders will depend on the ability to limit the expression of inserted globin genes to the appropriate tissues.[14]

Primate Studies. In early 1985, researchers from the National Institutes of Health and Memorial Sloan Kettering Cancer Center set out to develop a protocol for gene transfer that could, at the appropriate time, be applied to human beings. Their primate studies confirm that adequate numbers of bone marrow cells can survive manipulation outside the body to provide a large animal with a fully reconstituted blood-producing system. Like the mouse studies, however, they show that much remains to be learned before gene therapy will be feasible in a clinical setting.

The retroviral vector selected by the NIH investigators and their coworkers for the primate studies is called SAX. It is derived from the N2 vector and contains both the *neo* gene and a cDNA copy of the human *ADA* gene. Expression of the human *ADA* gene is regulated by a promoter taken from the monkey virus SV40.[15]

The first task for the NIH investigators and their coworkers was to develop an effective infection protocol for the large number of cells required for gene transfer in the monkeys. Initially, they tried cocultivating the monkey bone marrow cells with helper cells producing the SAX virus, because cocultivation for 24 hours is the most effective way to infect mouse bone marrow cells with the N2 vector. They found, however, that cocultivation was associated with loss of bone marrow cells—the total number of cells available for transplantation was too low to reconstitute a monkey's blood-producing system. Much better results were achieved by culturing the monkey bone marrow cells with virus-containing fluid harvested from the helper cell cultures. Using this latter protocol, four monkeys achieved full reconstitution.

Studies of infection levels and gene expression in the transplanted monkeys are more difficult to interpret. Analyses of blood cells from the monkeys have never revealed any vector DNA, but very low levels of human ADA were produced transiently by four out of five animals available for study (one monkey partially reconstituted by the cocultivation procedure and the four fully reconstituted animals).

The maximum level of human ADA activity found in the four monkeys was 0.5 percent of the level of monkey ADA activity. The researchers believe that this figure represents normal expression of human ADA by a few cells, rather than very low expression by many cells. Analyses focusing on the *neo* gene in the SAX construct indicate that expression of the vector genes occurred primarily in one cell type, T lymphocytes. The researchers do not have an explanation for this finding.

The most discouraging aspect of the monkey studies is that human ADA expression dropped to zero within five months of transplantation, although G418-resistant T cells (cells expressing the *neo* gene) could be obtained as late as seven months after treatment in one animal. The researchers offer many possible explanations for the transient expression of the vector-derived genes:

- Infection might have been achieved only in a small number of stem cells; because these cells did not have a selective advantage over other stem cells, their offspring were eventually diluted out.
- Infection might have been restricted to more mature bone marrow cells with a limited life span.

- Something about the infection procedure might have changed or damaged affected cells, leading to their removal by the immune system (for example, infected cells may undergo membrane modifications that mark them as foreign).
- A mechanism may exist to turn off the expression of foreign genes in individual cells over time.

The first situation would be the least problematic for two reasons. First, the low level of infection may be a technical problem that can be solved relatively easily. Second, while stem cells expressing the human *ADA* gene do not have a selective advantage in normal monkeys, they might have such an advantage in patients with ADA deficiency.

Overall, however, the large number of uncertainties associated with this study underscores the fact that researchers still do not understand the mechanisms that control vector expression in intact animals.

Safety

None of the mice or monkeys that have achieved full reconstitution of their blood-forming systems after gene transfer into bone marrow cells have shown any ill effects from the procedure. This finding is important, because many persons both inside and outside the scientific community have expressed concern about the potential risks of inserting DNA randomly into the cells of live animals. The consensus among researchers in the field is that the potential risks of gene therapy are small relative to the potential therapeutic benefits, and that new techniques in retrovirology and other fields will reduce these risks even further.

The safety issues in gene transfer can be divided into three categories. The first category concerns the purity of the biological materials used to produce the retroviral vectors. Researchers have many different techniques for identifying and eliminating possible contaminants in cell culture, so this category does not present any new or unusual problems. The second and third categories are more specific to gene transfer. They are the possibility of producing infectious virus particles and the possibility of causing harmful mutations in infected cells.

Ongoing Infection with a Vector Virus

Successful gene transfer with retroviruses is based on the concept that all virus particles produced by a helper cell should be capable of only one infection. After entering a target cell and integrating into the target cell genome, the vector provirus should be trapped (because it lacks the genes necessary to produce viral proteins; see Chapter 5). Transduced cells should not produce any intact retroviruses.

One of the potential disadvantages of using retroviruses as vectors in gene therapy is that they have the capability to exchange sequences with other retroviruses. This process, called **recombination,** could result in the development of an intact retrovirus. Infection with such a retrovirus would be unlikely to cause disease, but it could result in an inappropriate distribution of transferred genes inside the patient.

The most obvious concern is that the vector virus, which lacks the genes required to make essential viral proteins, might recombine with the helper virus (see Figure 6.8). The product

FIGURE 6.8   The helper cell shown here contains three proviruses: the helper provirus that enables it to make viral proteins, the vector provirus (carrying a human gene), and a defective endogenous provirus. Ideally, the helper cell should produce virus particles capable of only one infection (such a particle would consist of the vector RNA genome encapsulated in a protein core produced by the helper provirus). However, a virus capable of an ongoing infection could be generated through a recombination event. If a copy of the helper virus RNA genome were accidentally packaged with a copy of the vector virus RNA genome in a single virus particle, they might exchange bits of genetic information. (The same event could occur with the endogenous virus.) The result could be a target cell infected with a recombined intact provirus, capable of making all the proteins and RNA required to assemble an infectious virus particle. The text describes measures taken by scientists to minimize the potential for recombination events.

GENERATING A VIRUS CAPABLE OF AN ONGOING INFECTION VIA RECOMBINATION

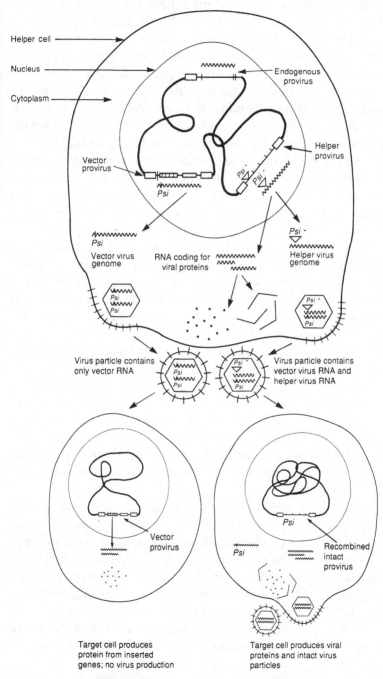

Helper cell

Nucleus

Cytoplasm

Endogenous provirus

Helper provirus

Vector provirus

Psi

Psi -

Psi -

Vector virus genome

Psi

RNA coding for viral proteins

Helper virus genome

Psi -

Psi

Psi

Psi

Psi -

Psi

Virus particle contains only vector RNA

Psi

Psi

Virus particle contains vector virus RNA and helper virus RNA

Psi -

Psi

Vector provirus

Psi

Psi

Recombined intact provirus

Target cell produces protein from inserted genes; no virus production

Desired outcome

Target cell produces viral proteins and intact virus particles

Recombination

of the recombination event might have the capability to produce all of the proteins and RNA required to assemble an infectious virus particle. Another possibility is that the vector virus could recombine with an endogenous provirus in the helper cell. Many animals carry copies of defective proviruses (representing viral infections in past generations) integrated into their genomes. Usually, the endogenous proviruses lack elements essential for viral replication, but recombination with a vector virus could restore them. The final possibility is that the vector could recombine with an endogenous provirus in the human target cell. No one knows how such a recombinant would behave, but if the endogenous provirus were derived from a cancer virus, the new virus might cause cancer.

This third type of recombination event is particularly unlikely, because recombination requires that the two viruses be closely related (have sizable DNA sequences in common). The retroviral vectors now in use are derived from mouse retroviruses—although mouse retroviruses and known human retroviruses have some sequences in common, their relationship is considered to be quite distant.

Several different measures are being taken to minimize the potential for recombination events in the helper cell. Scientists developing new vector systems attempt to restrict the similarities between vector viruses and helper viruses, and between vector viruses and known endogenous viruses. Also, researchers have designed new types of helper cells in which the helper virus has multiple alterations in addition to deletion of the packaging signal (the *Psi* sequence described in Chapter 5).

The final method for reducing the possibility of an ongoing infection with recombinant viruses involves the use of assays for virus production. Every vector engineered for gene transfer should be tested extensively in cultured human bone marrow cells (or other designated target cells) before it is used in patients. The cultured cells should be examined regularly for evidence of virus production.

Studies from other fields indicate that even if the above measures failed, widespread infection with a recombinant retrovirus would be very unlikely. Both monkeys and human beings appear to have immune mechanisms that rapidly inactivate mouse retroviruses.

Mutations

The random manner in which a retrovirus inserts its DNA into the genome of an infected cell can create two types of problems for the cell: it can inactivate a beneficial gene, or it can activate a harmful one. The second problem is more serious than the first, because its impact may extend beyond the individual cell to the entire organism.

Inactivation. Insertion of a provirus into the middle of a functional gene prevents the gene from being translated into protein (see Figure 6.9). Because cells usually have two copies of each gene, one on each of two paired chromosomes, inactivating one copy rarely has any effect. The normal gene can produce enough protein to maintain normal cell function. Occasionally, however, a cell may not have a second normal copy of a gene. In that case, insertion of the provirus may kill the cell.

Harold Varmus, a virologist from the University of California at San Francisco, estimates that the likelihood of an inactivation mutation is very small — about 1 in 100,000 for each cell infected. The problem does not create any potential risks for somatic cell gene therapy patients, because the loss of a few cells in a bone marrow graft would not affect the outcome of a gene transfer procedure.

Activation of Harmful Genes. A more serious concern for researchers in gene therapy is the potential for activating **proto-oncogenes.** Under most conditions, proto-oncogenes act like other genes—they carry the blueprints for proteins that participate in normal cell functions. The unique feature of proto-oncogenes is their association with cancer. Inappro-

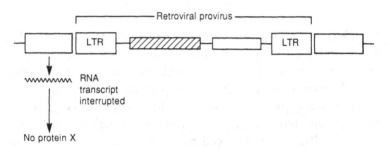

INACTIVATION MUTATION

Before Insertion

Gene X

RNA

Protein X

After Insertion

Retroviral provirus

LTR   LTR

RNA
transcript
interrupted

No protein X

FIGURE 6.9 Insertion of a retroviral provirus into the middle of a functional gene prevents the gene from being translated into protein. If the cell has a second copy of the disrupted gene, it can produce enough protein to maintain normal cell function. If the cell does not have a second copy of the gene, insertion of the provirus may kill the cell. This situation does not create any potential risks for somatic cell gene therapy patients, because the loss of a few cells in a bone marrow graft would not affect the outcome of a gene transfer procedure.

priate activation or alteration of a proto-oncogene turns it into an **oncogene,** a powerful inducer of unregulated growth.

Experiments in laboratory animals indicate that the insertion of a retroviral provirus next to a proto-oncogene can provide the stimulus necessary for cancerous transformation. Regulatory elements in the viral LTR may activate the expression or enhance the level of expression of the harmful gene (see Figure 6.10).

ACTIVATION OF HARMFUL GENES

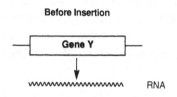

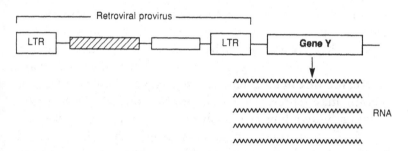

FIGURE 6.10 Animal studies indicate that insertion of a retroviral provirus next to a proto-oncogene (a gene capable of becoming an oncogene—a powerful inducer of unregulated growth) can provide the stimulus necessary for cancerous transformation. Before insertion of the provirus, expression of the proto-oncogene Y is very limited. After insertion, elements in the viral LTR enhance the level of expression of the harmful gene. For reasons described in the text, researchers in the field of gene therapy believe that the risk of this type of event occurring in gene therapy patients is low.

Researchers in the field of gene therapy believe that the risk of this type of event occurring in gene therapy patients is low for several reasons. First, the number of proto-oncogenes in the human genome is very small compared with the total number of genes in a cell.[16] Thus, the likelihood that a provirus will be inserted next to a proto-oncogene in a single round of infection is small.[17] Second, the activity of proto-oncogenes appears to be cell-specific; a proto-oncogene capable of transforming a liver cell might be totally inactive in a bone marrow cell. Third, not every virus is efficient at turning on proto-oncogenes. Finally, current evidence suggests that tumor formation probably requires more than one type of mutation—the insertion of a retrovirus next to a proto-oncogene may not be enough to cause cancer by itself. The development of self-inactivating vectors, described above, may reduce the risk of cancer in gene therapy patients even further.[18]

Comparative Safety of Gene Therapy

The small risk of cancer associated with human somatic cell gene therapy has alarmed some observers, but physicians emphasize that the risk must be assessed in a clinical context. The first candidates for gene therapy will be patients who have severe, life-threatening genetic diseases. For these people, gene therapy may represent the only hope of surviving early childhood or of avoiding progressive mental and physical deterioration. Gene therapy researchers believe that the potential benefits of an effective and reliable gene therapy protocol will outweigh the potential risks in many cases.

Patients with other conditions often face similar issues. For example, the recipients of solid organ transplants usually require life-long immune suppression to prevent graft rejection. In a certain proportion of these patients, the immunosuppressive therapy leads to cancer. Similarly, the chemotherapy or radiation therapy used to treat a childhood malignancy may lead to the development of a different malignancy many years later.

Bone marrow transplantation has saved many lives, but it too has important limitations (see Chapter 4). Between 10 and 20 percent of bone marrow transplant patients succumb to graft versus host disease and related complications within a year. Also, prospective recipients often must be prepared with high doses of chemotherapy to reduce the risk of graft rejection and to provide physiological space for donor cells.

The weighing of potential risks and benefits for the first gene therapy patients will require a careful evaluation of data from cell culture studies and from animal experiments. Researchers will have to show that the procedures they plan to use have a high probability of benefiting the patient and that appropriate safety tests have been conducted.

Conclusions

Recent gene transfer experiments have given scientists a new appreciation of the challenges they face in developing effective procedures for gene therapy in human beings. Animal studies have shown that successful expression of an inserted gene depends on many factors: the origin and structure of the vector; the properties of the inserted gene; the site of integration of the foreign DNA; and the nature and developmental status of the target cell.

Researchers have only begun to explore the range of possible vector formats. Vectors with internal promoters and genomic vectors appear to offer the most hope for the future, but many more must be tested to understand why some succeed and others fail. Data showing changes in the expression of inserted genes over time underscore the need for more emphasis on long-term studies in live animals.

Another clear message derived from existing research is that scientists cannot assume that a vector structure that results in the expression of one human gene will function the same way when a different gene is inserted. This creates special problems for researchers attempting to develop a rational approach to clinical trials in human beings. Each vector considered for human use must be tested separately; it simply is not possible to extrapolate from one vector to another.

The ideal culture system in which to test vectors intended for use in human beings would be a system consisting of primitive human stem cells. But researchers still cannot identify these cells or separate them from other constituents of the bone marrow. Widespread application of gene therapy probably will require a better understanding of the factors that control blood cell development in laboratory animals and in human beings.

The consensus among scientists in the field of gene therapy is that recent studies have provided valuable new information on which to base future research efforts, but they have not satisfied the criteria listed at the beginning of this chapter as prerequisites for human trials. Gene therapy will not become a reasonable alternative for patient care until researchers can achieve higher and more stable levels of expression of inserted genes in laboratory animals.

Most scientists are unwilling to predict when these goals will be met—estimates vary from two years to more than ten years. The answers depend in part on individual views about medical progress. Throughout history, certain forms of therapy have been used long before physicians understood how they worked in the human body (examples include digitalis, aspirin, and the smallpox vaccine). Local and national review boards, described in Chapter 8, will have to decide how much information is necessary before researchers can use their limited knowledge of gene transfer to attempt to help patients with life-threatening genetic diseases.

—— ACKNOWLEDGMENTS ——

Chapter 6 is based on the presentations of W. French Anderson, Mario R. Capecchi, Robert J. Desnick, Theodore Friedmann, Eli Gilboa, David W. Martin, Jr., A. Dusty Miller, Richard Mulligan, Stuart Orkin, Harold E. Varmus, and Inder Verma.

—— NOTES ——

1. Anderson, W. F., and J. C. Fletcher. "Gene Therapy in Human Beings: When Is It Ethical to Begin?" *New England Journal of Medicine* 303:1293-1297.

2. For descriptions of these diseases, see Chapter 2.

3. Lesch-Nyhan syndrome is an X-linked disorder in which

affected males lack the enzyme hypoxanthine-guanine phosphoribosyltransferase (HPRT).

4. In the future, animal models also may be developed using gene targeting (inducing newly introduced DNA sequences to recombine with corresponding DNA sequences residing in the chromosome). Mario Capecchi, an authority on gene targeting from the University of Utah, says that researchers will soon have the ability to disable specific genes in the genome of a laboratory animal.

5. Interpreting the results of gene transfer studies in animal models may be particularly difficult. A retroviral vector constructed from a virus that normally infects mice might behave very differently in human beings than it does in the mouse.

6. The vast majority of gene transfer studies in animals have employed bone marrow cells. Work also is in progress using skin cells and liver cells (see Chapter 5).

7. The *neo* and *DHFR* genes are selectable markers (see Chapter 5). They are not genes that would be used to treat a genetic disease.

8. The researchers traced the ancestry of mature blood cells in the transplanted mice by examining their DNA. The site of insertion of the retroviral provirus in the cell genome was used as a marker to identify cells with a common ancestor. The discovery of lymphoid cells and macrophages with identical insertion patterns demonstrated that some of the transduced bone marrow cells were true stem cells—they had given rise to the entire spectrum of blood cells in the body.

9. Most studies of gene transfer have focused on short-term results. In one common assay, researchers examine spleen cells from mice two weeks after they have received bone marrow transplants containing transduced cells. At that time, virtually all of the cells in the spleen are derived from the transplanted marrow (because the mice are irradiated before the transplant to provide physiological space for the treated cells). By analyzing DNA extracted from spleen cell colonies, researchers determine the proportion of bone marrow cells in the original graft that incorporated the retroviral vector.

10. The role of the helper or packaging cell in the construction of retroviral vectors is described in Chapter 5.

11. Scientists design a retroviral vector that has one intact LTR and one LTR from which the promoter and enhancer sequences have been deleted. When the vector enters a target cell and is transcribed back into DNA (see Chapter 5 for a description of the life cycle of a retrovirus), the deletion is transferred to both LTRs. Thus, the

provirus in the target cell does not have an active transcriptional unit to intefere with the expression of inserted foreign genes.

12. Researchers at the National Heart, Lung, and Blood Institute, NIH, and at MIT have achieved good results in laboratory studies of genomic vectors carrying whole human globin genes, but other investigators have been less successful.

13. Immediately after transplantation, blood cells taken from transplanted animals survive and multiply in a culture medium containing G418. Within a short period of time, however, the ability of these cells to form colonies in the selective medium disappears.

14. Scientists do not have sufficient knowledge about gene function to anticipate the problems that might arise in a patient if the beta-globin gene were expressed for long periods of time in inappropriate tissues (for example, white blood cell populations).

15. In another study by the same researchers, infection with the SAX vector resulted in normal ADA expression by human cells that previously had been unable to produce the enzyme. (The cultured cells were derived from ADA-deficient T lymphocytes.)

16. Researchers have identified more than 30 distinct proto-oncogenes. The actual number of proto-oncogenes in a human cell may be between 100 and 200. The total number of genes in a human cell is about 100,000.

17. The risk of harmful mutations would increase if the vector virus were capable of replicating itself and producing an ongoing viral infection.

18. The absence of enhancer and promoter sequences in the two LTRs of the integrated self-inactivating vector may reduce the possibility of activating cellular oncogenes located near the provirus.

──── SUGGESTED READINGS ────

Anderson, W. French, and John C. Fletcher. 1980. "Gene Therapy in Human Beings: When Is It Ethical to Begin?" 1980. *New England Journal of Medicine* 303:1293-1297.

Animal Models of Human Disease. 1972-present (looseleaf). Registry of Comparative Pathology. Washington, D.C.: Armed Forces Institute of Pathology.

Costantini, Frank, Kiran Chada, and Jeanne Magram. 1986. "Correction of Murine β-Thalassemia by Gene Transfer into the Germ Line." *Science* 233:1192-1194.

Desnick, Robert J., Donald F. Patterson, and Dante G. Scarpelli, eds. 1982. *Animal Models of Inherited Metabolic Diseases.* New York: Alan R. Liss.

Dick, John E., Maria Cristina Magli, Dennis Huszar, Robert A. Phillips, and Alan Bernstein. 1985. "Introduction of a Selectable Gene into Primitive Stem Cells Capable of Long-Term Reconstitution of the Hemopoietic System of W/W^v Mice." *Cell* 42:71-79.

Eglitis, Martin A., Philip Kantoff, Eli Gilboa, and W. French Anderson. 1985. "Gene Expression in Mice after High Efficiency Retroviral-Mediated Gene Transfer." *Science* 230:1395-1398.

Gilboa, Eli. 1987. "Retrovirus Vectors and Their Uses in Molecular Biology." *BioEssays* 5:252-257.

Gilboa, Eli, Martin A. Eglitis, Philip W. Kantoff, and W. French Anderson. 1986. "Transfer and Expression of Cloned Genes Using Retroviral Vectors." *BioTechniques* 4(6):504-512.

Hock, Randy A., and A. Dusty Miller. 1986. "Retrovirus-Mediated Transfer and Expression of Drug Resistance Genes in Human Haematopoietic Progenitor Cells." *Nature* 320:275-277.

Kantoff, Philip W., Al Gillio, Jeanne R. McLachlin, Claudio Bordignon, Martin A. Eglitis, Nancy A. Kernan, Robert C. Moen, Donald B. Kohn, Sheau-Fung Yu, Evelyn Karson, Stefan Karlsson, James A. Zwiebel, Eli Gilboa, R. Michael Blaese, Arthur Nienhuis, Richard J. O'Reilly, and W. French Anderson. 1987 "Expression of Human Adenosine Deaminase in Non-Human Primates after Retroviral-Mediated Gene Transfer." *Journal of Experimental Medicine* 166:219-234.

Kantoff, Philip W., Donald B. Kohn, Hiroaki Mitsuya, Donna Armentano, Miri Sieberg, James A. Zwiebel, Martin A. Eglitis, Jeanne R. McLachlin, Dan A. Wiginton, John J. Hutton, Sheldon D. Horowitz, Eli Gilboa, R. Michael Blaese, and W. French Anderson. 1986. "Correction of Adenosine Deaminase Deficiency in Cultured Human T and B Cells by Retrovirus-Mediated Gene Transfer." *Proceedings of the National Academy of Sciences* 83:6563-6567.

Karlsson, Stefan, Thalia Papayannopoulou, Stefan G. Schweiger, George Stamatoyannopoulos, and Arthur W. Nienhuis. 1987. "Retroviral-Mediated Transfer of Genomic Globin Genes Leads to Regulated Production of RNA and Protein." *Proceedings of the National Academy of Sciences* 84:2411-2415.

Keller, Gordon, Christopher Paige, Eli Gilboa, and Erwin F. Wagner.

1985. "Expression of a Foreign Gene in Myeloid and Lymphoid Cells Derived from Multipotent Haematopoietic Precursors." *Nature* 318:149-154.

Kolata, Gina. "Gene Therapy Method Shows Promise." 1984. *Science* 223:1376-1379.

Kuehn, Michael R., Allan Bradley, Elizabeth J. Robertson, and Martin J. Evans. 1987. "A Potential Animal Model for Lesch-Nyhan Syndrome through Introduction of HPRT Mutations into Mice." *Nature* 326:295-298.

Kwok, William W., Friedrich Schuening, Richard B. Stead, and A. Dusty Miller. 1986. "Retroviral Transfer of Genes into Canine Hemopoietic Progenitor Cells in Culture: A Model for Human Gene Therapy." *Proceedings of the National Academy of Sciences* 83:4552-4555.

Lemischka, Ihor R., David H. Raulet, and Richard C. Mulligan. 1986. "Developmental Potential and Dynamic Behavior of Hematopoietic Stem Cells." *Cell* 45:917-927.

Marx, Jean L. 1986. "Gene Therapy—So Near and Yet So Far Away." *Science* 232:824-825.

——— 1987. "Probing Gene Action during Development: The Good News—and the Bad—about Gene Therapy Prospects." *Science* 236:29-30.

Merz, Beverly. 1986. "Stumbling Blocks Pave Path to Clinical Trials for Gene Therapy." *JAMA* 255(14):1825-1827, 1832.

Readhead, Carol, Brian Popko, Naoki Takahashi, H. David Shine, Raul A. Saavedra, Richard L. Sidman, and Leroy Hood. 1987. "Expression of a Myelin Basic Protein Gene in Transgenic Shiverer Mice: Correction of the Dysmyelinating Phenotype." *Cell* 48:703-712.

Stead, Richard B., William W. Kwok, A. Dusty Miller, and Rainer Storb. 1986. "A Canine Model for Gene Therapy: Lack of Efficient Gene Expression or In Vivo Selection in Dogs Reconstituted with Autologous Marrow Infected with Retroviral Vectors." *Blood* 68:307A (abstract).

Thomas, Kirk R., and Mario R. Capecchi. 1986. "Introduction of Homologous DNA Sequences into Mammalian Cells Induces Mutations in the Cognate Gene." *Nature* 324:34-38.

Thomas, Kirk R., Kim R. Folger, and Mario R. Capecchi. 1986. "High Frequency Targeting of Genes to Specific Sites in the Mammalian Genome." *Cell* 44:419-428.

Williams, David A., Stuart H. Orkin, and Richard C. Mulligan. 1986. "Retrovirus-Mediated Transfer of Human Adenosine Deaminase Gene Sequences into Cells in Culture and into Murine Hematopoietic Cells In Vivo." *Proceedings of the National Academy of Sciences* 83: 2566-2570.

7 | Ethical and Economic Issues

Gene splicing is a revolutionary scientific technique that recasts past ideas and reshapes future directions. Even so, it does not necessarily follow that all its applications or objectives represent a radical departure from the past.

PRESIDENT'S COMMISSION FOR THE STUDY
OF ETHICAL PROBLEMS IN MEDICINE AND
BIOMEDICAL AND BEHAVIORAL RESEARCH

In June 1980, the general secretaries of three large religious bodies in the United States—the U.S. Catholic Conference, the Synagogue Council of America, and the National Council of Churches—sent a letter to President Carter in which they expressed concern about the rapid pace of advances in the new field of genetic engineering. The letter reflected their belief that genetic engineering represented a radical departure from previous technologies and that no government agency or committee had addressed the fundamental ethical and moral issues raised by the ability to manipulate human genes.

Several months later, the President's Commission for the Study of Ethical Problems in Medicine and Biomedical and Behavioral Research began a two-year study that culminated in the publication of *Splicing Life: The Social and Ethical Issues of Genetic Engineering with Human Beings.*[1] In this comprehensive document, the Commission sought to clarify concerns about genetic engineering and to develop a foundation for public policy in the areas of genetically engineered medical products, genetic screening and diagnosis, and gene therapy.

Addressing Public Concerns

The Commission noted the wide range of popular views on gene therapy. Some persons hail the technique as a weapon for fighting every genetic disease known. Others worry that one step into the arena of gene therapy will lead inexorably to practices that reduce the human species to "a technologically designed product."[2] Clearly, the expectations of the first group are overly optimistic. Scientists are only beginning to understand how to manipulate a single gene; for the foreseeable future, gene therapy will be limited to a small number of inherited disorders that result from the absence or inactivity of a single gene product.

The fear that gene therapy might "reduce" the human species or "change the meaning of being human" raises many different issues. First, it presupposes that gene therapy differs in significant ways from accepted forms of treatment. In fact, somatic cell gene therapy—the only form of gene therapy now being considered for clinical trials—represents a logical extension of existing approaches. The President's Commission concluded:

> Gene therapy carried out on somatic cells, such as bone marrow cells, would resemble standard medical therapies in that they all involve changes limited to the cells of the person being treated. They differ, however, in that gene therapy involves an inherent and probably permanent change in the body rather than requiring repeated applications of an outside force or substance. An analogy is organ transplantation, which also involves the incorporation into an individual of cells containing DNA of 'foreign' origin.[3]

Bone marrow transplantation and the transplantation of solid organs are standard medical practices for treating many disorders. Clinical trials examining new applications of these techniques are judged by the same criteria applied to other forms of human therapeutic experimentation (based on international guidelines for the protection of human subjects). Neither bone marrow transplantation nor somatic cell gene therapy raises any special ethical issues.

Germ Line Gene Therapy

Germ line gene therapy is a different matter, because it constitutes a departure from previous medical interventions. In germ line gene therapy, genetic changes would be passed on deliberately to future generations. For now, technical uncertainties associated with the procedure (the high failure rate in animal models, the risk of a deleterious result, and the lack of techniques to assess the genetic status of embryos) preclude consideration of germ line gene therapy for human use. If and when these problems are solved, acceptance of the procedure will depend on the resolution of a range of complex ethical issues.

LeRoy Walters, director of the Center for Bioethics at Georgetown University, explains that there would be two basic rationales for using germ line gene therapy to treat genetic diseases. The first would arise if some genetic diseases were found to be resistant to somatic cell gene therapy. For example, scientists might find that the blood–brain barrier prevented the insertion of normal genes into nerve cells affected by hereditary disorders of the central nervous system. The only way to get around the barrier might be to insert genes into the cells of a very early embryo, using a technique that did not differentiate between somatic cells and germ cells.

The second rationale for germ line gene therapy, according to Walters, would be one of efficiency. If somatic cell gene therapy became a successful cure for some of the more prevalent genetic diseases, such as cystic fibrosis, the treated patients would constitute a new group of homozygous "carriers." If two treated patients with the same genetic abnormality had children, all of their offspring would be affected by the disease. Although succeeding generations of affected children presumably could be treated with somatic cell gene therapy, patients and their physicians might view germ line gene therapy as a more efficient alternative, because at least some offspring would inherit the inserted gene from their parents.[4]

The ethical arguments against the use of germ line gene therapy to treat genetic diseases also fall into two categories. The first relates to the potential risks of gene therapy: for example, is it ethical to use a technique in which unanticipated problems or mistakes could be passed on to future generations? The second category consists of broader concerns about changes to the gene pool—the genetic inheritance of the human population. The congressional Office of Technology Assessment addressed the latter category in its report *Human Gene Therapy: A Background Paper*:

> Direct manipulation of the genome inspires visions of mankind controlling its own evolution, depleting the diversity of genes in the human population, and crossing species barriers to create new life forms. The magnitude and rapidity of change caused by direct genetic intervention, however, are likely to be far smaller than the large effects caused by relaxing historic selection pressures on the human population through changes in the environment, sanitation, and health care.[5]

The OTA analysts determined that direct germ line gene therapy of recessive disorders would have a noticeable effect on human evolution only if widely practiced for hundreds of generations. They noted that arguments for refraining from gene therapy to maintain genetic diversity are weakened by the tendency for genetic diversity to increase in any rapidly expanding population. Finally, they concluded,

> Even if gene therapy *did* have an effect on genetic diversity, this might not prevent its use. The risk of slightly reducing diversity in the entire human population would likely seem insignificant to those patients for whom the potential benefits loom large and immediate. Perpetuation of genetic disease, particularly of the severe childhood diseases that are now the targets for gene therapy, would seem a cruel means to an end of uncertain import.[6]

The President's Commission, the authors of the OTA report, and scientists involved in the development of gene therapy techniques have emphasized the importance of widespread public discussion of human germ line gene therapy before it becomes technically feasible. Scientists, educators, public officials, and the media will have to work together to provide the educational framework in which such a discussion can take place.

Altering "Normal" Individuals

The underlying concern among those who object most strenuously to the use of human gene therapy is that the new techniques eventually might be employed to alter genetic traits in "normal" individuals. The prospect of parents turning to genetic engineering as a way of producing "perfect" children is extremely disturbing. Such behavior would threaten basic concepts about the value of the individual in our society. As explained in Chapter 1, however, eugenic genetic engineering is beyond the realm of scientific possibility at this time. Traits such as intelligence, personality, and the formation of body parts involve many different genes interacting in unknown ways with the environment; they could not be altered by available techniques.

The Human Gene Therapy Subcommittee of the NIH Recombinant DNA Advisory Committee notes that research in molecular biology eventually could lead to the development of genetic techniques to enhance human capabilities, but it "does not believe that these effects will follow immediately or inevitably from experiments with somatic cell gene therapy."[7]

Diverse religious, scientific, and governmental organizations have concluded that it would be unethical to withhold somatic cell gene therapy from severely ill patients—assuming it can be shown to be both safe and effective—solely because other forms of genetic engineering might be misused in the future.

Clinical Research

Human somatic cell gene therapy does not raise any new ethical problems; however, researchers must address ethical issues common to the development of all innovative therapies. The need to demonstrate that the potential benefits of a new technique outweigh the potential risks is discussed in Chapter 6. Nontechnical ethical issues can be divided into three areas: patient selection, informed consent, and privacy and confidentiality.

Patient Selection

Medical researchers in several fields have described the difficulties associated with selecting patients for early clinical trials of new, life-saving procedures. If the number of potential subjects exceeds the number of places available in the study, it may be very difficult to avoid making judgments about comparative social worth.

Two factors limit the scope of this problem in gene therapy for genetic diseases. The first is the rarity of genetic diseases considered to be candidates for early gene therapy trials. The second is the young age of the prospective patients—most will be in early childhood (when issues of comparative social worth are less likely to arise). Nonetheless, genetic disease specialists must be aware of the potential problem and develop procedures to handle it should the need arise.[8]

Gene Therapy in Cancer Treatment.
So far, this book has focused exclusively on the treatment of genetic diseases, but increasingly researchers are exploring the possibility of using somatic cell gene therapy in cancer treatment. Two basic strategies have been considered. The first would involve inserting genes into normal bone marrow cells to make them more resistant to toxic, anti-cancer drugs; the second would involve inserting genes into cancer cells to make them more susceptible to chemotherapy.

The selection of patients for early clinical trials of these procedures may be more difficult than the selection of patients for other types of gene therapy studies, because the number of potential subjects would be much greater.

Fetal Therapy. If somatic cell gene therapy is found to be useful in the treatment of severe diseases of childhood, there will be a strong incentive to develop techniques for gene therapy in fetuses. Researchers have extensive evidence that the metabolic abnormalities associated with some genetic diseases cause serious damage before birth. For example, some patients with galactosemia are born with cataracts and others exhibit speech defects and learning disabilities despite early placement on a galactose-free diet. Similarly, studies of brain tissue from fetuses with Tay-Sachs disease have revealed early signs of nerve cell damage. The goal of fetal gene therapy would be to intervene before permanent damage occurred.

The major roadblock to the development of gene therapy for human fetuses (assuming that somatic cell gene therapy proves to be safe and effective in infants and young children) will be widespread prohibitions against fetal research. Scientists planning strategies for intrauterine gene therapy will need extensive information about gene expression during normal human development. Many believe that the only way to obtain such information will be through fetal research.

The United States has had a legislative ban on federally supported fetal research since November 1985 (effective until October 31, 1988).[9] In addition, 25 states have statutes that limit or prohibit such research. Thus, the application of gene therapy to fetuses may require studies in countries where fetal research is practiced, or a reevaluation of current fetal research guidelines in the United States.

Informed Consent

All researchers planning clinical trials in the United States must follow federally mandated procedures to obtain the informed consent of prospective research subjects.[10] Several factors will make this process particularly difficult for early trials of gene therapy. They include the age of prospective patients, the newness of the approach, and the complex character of the target diseases.

Most scientists anticipate that human somatic cell gene therapy will be used initially to treat diseases that are fatal early in life. Therefore, the subjects of early clinical trials will be infants and children who are too young to understand the potential risks and benefits involved. Decisions about participation will fall on parents or guardians.

Scientists will have to work closely with these parents to provide them with basic information about bone marrow transplantation and the mechanics of gene transfer. The low but finite risks of genetic mutation, persistent retroviral infection, and other adverse effects must be weighed against the consequences of the disease. Above all, LeRoy Walters says, parents must understand that they are "embarking on essentially uncharted territory."

William Krivit, a professor in the Department of Pediatrics at the University of Minnesota Medical School, recently described the process of obtaining informed consent for clinical trials of bone marrow transplantation in patients with lysosomal storage disorders that affect the brain. He underscored the importance of helping parents understand that bone marrow transplantation might prolong the lives of their children (by halting or preventing damage to other organs in the body), but have no impact on neurological deterioration. (As explained in Chapter 4, researchers are not yet certain about the extent to which bone marrow transplantation can prevent or arrest brain damage in patients with diseases such as Hurler syndrome.)

Similar issues could arise in gene therapy trials if the diseases involved multiple organ systems in the body. To avoid unrealistic expectations, parents (and patients, if they are old enough) must have a thorough understanding of the range of possible outcomes before therapy begins.

Privacy and Confidentiality

An important part of obtaining consent from the families of prospective subjects for early gene therapy trials will be making them aware of the potential for widespread publicity surrounding the clinical trials. Current NIH guidelines require that proposals for human somatic cell gene therapy be published in abstract form in the *Federal Register* prior to consideration by the Recombinant DNA Advisory Committee (RAC). Meetings of the RAC will be open to the public and will include an opportunity for public comment. These activities will draw widespread attention from the media and from government, scientific, community, and religious leaders.

In such an environment, researchers probably will not be able to protect the identities of their subjects once treatment has begun. Institutions involved in early gene therapy trials will have to plan carefully to maintain a reasonable balance between familial privacy and the public's desire for detailed information.

The issues of privacy and confidentiality raised by gene therapy research are very similar to those raised by other novel forms of therapy. In the long run, they will be much easier to resolve than the complex questions raised by the expanding field of genetic diagnosis. As explained in Chapter 3, scientists will soon have the capability to identify genetic risk factors for some of the most common diseases in human beings. Society faces a major challenge in deciding how this predictive capability will be used.

Economic Issues

For most single-gene disorders, modern medicine and surgery have very little to offer. An effective therapy could have a very high one-time cost and still be more cost-effective than the years of repeated hospitalizations experienced by children with diseases such as severe combined immune deficiency and sickle cell disease.

One way to estimate the potential cost of gene therapy is to look at the cost of bone marrow transplantation. Robertson Parkman, head of the Division of Research Immunology and Bone Marrow Transplantation at Childrens Hospital of Los Angeles, says that bone marrow transplantation costs between $50,000 and $150,000. (The cost for an individual patient depends on the hospital, the complications that arise, and the patient's age. Children usually fare better than adults, so their costs are lower.)

The economic benefits of gene therapy would be most

striking for patients who would otherwise require long-term institutional care (for example, patients with one of the lysosomal storage disorders that causes deterioration of the central nervous system). Dr. Parkman estimates that the cost of hospitalization for such a child in California is now about $30,000 per year, with an expected life span of up to 20 years.

These figures suggest that successful gene therapy may become a cost-effective method of caring for children with many, if not most, single-gene defects, even without placing an economic value on improvements in the quality of patients' lives.[11] The ultimate impact of the procedure will depend, however, on how it compares with other forms of therapy at the time it becomes practicable. Advances in enzyme replacement therapy or bone marrow transplantation (for example, new solutions to the problem of graft versus host disease) might make gene therapy a relatively less desirable alternative in the future.

Conclusions

The ethical issues raised by human somatic cell gene therapy are the same as those raised by other new forms of therapy. Prior to clinical trials, researchers will have to demonstrate that the potential benefits of therapy are greater than the potential risks; that therapies known to be effective will not be withheld for purposes of conducting the trials; that the process of selecting patients will be fair; and that prospective subjects or their parents or guardians will be fully informed about all aspects of the procedure, including the fact that some adverse side effects may be irreversible.

The issues surrounding germ line gene therapy are more complex, because genetic changes would be passed on to future generations. Government, religious, civic, and scientific leaders should encourage widespread public discussion of the pros and cons of germ line gene therapy, even though it is unclear whether the approach will ever be technically feasible for human beings.

The ultimate clinical impact of somatic cell gene therapy will depend on how it compares with other forms of treatment.

At the moment, it appears that gene therapy may be a cost-effective method of treating many severely ill children. However, simultaneous advances in other areas of medicine could reduce its impact in the future.

—— ACKNOWLEDGMENTS ——

Chapter 7 is based on the presentations of W. French Anderson, Robertson Parkman, Leon E. Rosenberg, and LeRoy Walters.

—— NOTES ——

1. President's Commission for the Study of Ethical Problems in Medicine and Biomedical and Behavioral Research (Washington, D.C.: U.S. Government Printing Office, Stock no. 83-600500, 1982).

2. J. Rifkin, *Algeny* (New York: Viking, 1983), p. 233.

3. *Splicing Life*, p. 45.

4. It is important to distinguish between adding a normal gene to the genome and the much more difficult task of repairing or replacing an abnormal gene (which has not been accomplished in mammals). The random insertion of a normal gene into germ cells would not stop transmission of the corresponding abnormal gene to future generations. However, offspring who received the inserted gene would be protected from the effects of the genetic disease and could pass the normal gene on to some of their own children.

5. *Human Gene Therapy—A Background Paper* (Washington, D.C.: U.S. Congress, Office of Technology Assessment, OTA-BP-BA-32, 1984), p. 31.

6. *Human Gene Therapy—A Background Paper*, p. 32.

7. "Points to Consider in the Design and Submission of Human Somatic-Cell Gene Therapy Protocols." 1986. *Recombinant DNA Technical Bulletin* 9(4):221-242. (See Appendix.)

8. The issue of equitable access to care will become more prominent if somatic cell gene therapy passes through the experimental stage and becomes a standard part of medical practice (see Chapter 8).

9. There had been a de facto moratorium on research involving human fetuses for more than a decade before the 1985 legislation.

10. Code of Federal Regulations, Title 45, Part 46, March 8, 1983.

11. A thorough cost-effectiveness analysis would have to address an array of issues in addition to hospitalization costs.

—— SUGGESTED READINGS ——

Anderson, W. French. 1985. "Human Gene Therapy: Scientific and Ethical Considerations." *Journal of Medicine and Philosophy* 10:275-291.

Grobstein, Clifford, and Michael Flower. 1984. "Gene Therapy: Proceed with Caution." *Hastings Center Report* 14:13-17.

Human Gene Therapy—A Background Paper. 1984. Washington, D.C.: U.S. Congress, Office of Technology Assessment. OTA-BP-BA-32.

Krivit, William. 1986. "Conclusions: Ethics, Cost, and Future of BMT for Lysosomal Storage Diseases." In William Krivit and Natalie W. Paul, eds., *Bone Marrow Transplantation for Treatment of Lysosomal Storage Diseases,* pp. 189-194. March of Dimes Birth Defects Foundation, Birth Defects Original Article Series, vol. 22, no. 1. New York: Alan R. Liss.

Merz, Beverly. 1987. "Gene Therapy May Have Future Role in Cancer Treatment." *JAMA* 257(2):150-151.

Motulsky, Arno G. 1983. "Impact of Genetic Manipulation on Society and Medicine." *Science* 219:135-140.

President's Commission for the Study of Ethical Problems in Medicine and Biomedical and Behavioral Research. 1982. *Splicing Life: The Social and Ethical Issues of Genetic Engineering with Human Beings.* Washington, D.C.: U.S. Government Printing Office. Stock no. 83-600500.

Rosenberg, Leon E. 1985. "Can We Cure Genetic Disorders?" In Aubrey Milunsky and George J. Annas, eds., *Genetics and the Law III,* pp. 5-13. New York: Plenum Press.

Walters, LeRoy. 1986. "The Ethics of Human Gene Therapy." *Nature* 320:225-227.

8 | Federal Oversight of Gene Therapy

The publication of *Splicing Life* by the President's Commission for the Study of Ethical Problems in Medicine and Biomedical and Behavioral Research generated broad discussion within the federal government. The Commission's principal conclusions emphasized the need for a permanent oversight body to evaluate social and ethical issues, as well as technical considerations, raised by the application of genetic engineering to human beings. The Commission recommended that the oversight body have a diverse membership and that it regard education—of both the scientific community and the public—as one of its primary objectives. Several formats were suggested for the body, including a restructuring of the Recombinant DNA Advisory Committee (RAC).

The Recombinant DNA Advisory Committee

RAC was established within the National Institutes of Health (NIH) in 1974 to develop guidelines for the safe conduct of research involving recombinant DNA. Technically, RAC guidelines apply only to institutions that receive federal funding (the main penalty for violating the guidelines is withdrawal of those funds), but their influence has shaped the development of biotechnology worldwide. Private genetic engineering firms in the United States have submitted relevant research proposals to RAC voluntarily, and many foreign countries have used the guidelines as a foundation for their own policies on re-combinant DNA research.

Initially, RAC evaluations focused exclusively on health and environmental risks that might arise from the manipulation of DNA in the laboratory. As experience with recombinant DNA accumulated, however, the scope of RAC activities began to change. RAC guidelines were relaxed to reflect increased knowledge and the fact that many of the risks anticipated by those wary of genetic engineering experiments had not materialized. Local Institutional Biosafety Committees (IBCs) assumed much of the responsibility for assessing the safety of individual research projects. RAC began devoting more time to issues raised by the impending commercial development of the new technology.

Throughout its early years, critics of RAC expressed concern that the committee's membership was too narrow. The first members all had scientific or technical backgrounds, and all but one were affiliated with major research institutions. Two nonscientists were appointed in 1976,[1] but the balance of the committee did not change substantially until late 1978. In December of that year, Joseph Califano, then Secretary of the Department of Health, Education, and Welfare, reorganized RAC by enlarging and broadening its membership. The new committee was almost twice as big as its predecessor, and one-third of its 25 members were nonscientists.

Like the original, the new RAC dealt primarily with issues arising from the pharmaceutical and agricultural applications of recombinant DNA technology. Before the 1982 publication of *Splicing Life*, there was very little discussion of the technical, social, and ethical issues surrounding human gene therapy.

Response to the President's Commission

In April 1983, RAC decided to establish a special working group to evaluate the options presented in *Splicing Life*. The working group agreed with the President's Commission that RAC's approach to the application of genetic technology to human beings should be broadened. It recommended that:

(1) The membership of the RAC be modified to include adequate representation to deal credibly with these [social and ethical] issues.

(2) Procedures should be developed for the coordinate consideration of experiments involving the use of recombinant DNA technology in humans by Institutional Review Boards (IRBs), the Office for Protection from Research Risks (OPRR), the Food and Drug Administration (FDA), Institutional Biosafety Committees (IBCs), the Office of Recombinant DNA Activities (ORDA), and the Recombinant DNA Advisory Committee.

(3) The NIH Guidelines for Research Involving Recombinant DNA Molecules should be reviewed for their adequacy and clarity in dealing with human experimentation.[2]

The working group concluded that the appropriate role for RAC would be to review proposals for gene therapy on a case-by-case basis, after they had been approved by local Institutional Review Boards (whose function is to determine whether proposals meet federal regulations for the protection of human subjects) and Institutional Biosafety Committees. The working group noted that IRBs were required by law to focus on the interests of individual research subjects, and that no other mechanism existed for "evaluating the effect on the broader community of procedures involving use of recombinant DNA in humans."[3]

RAC accepted the recommendations of the working group and, after further deliberations, established the permanent Subcommittee on Human Gene Therapy. The subcommittee membership now includes three laboratory scientists, three clinicians, three ethicists, three attorneys, two specialists in public policy, and a representative of the lay public.

Evaluating Gene Therapy Proposals

The subcommittee's first task was to develop a framework for evaluating human gene therapy proposals. The format chosen by the group was a "Points to Consider" document, a set of technical and ethical questions designed to provide guidance for investigators preparing proposals for clinical trials. The document was published twice in the *Federal Register* for public review, modified in response to comments, and formally accepted by RAC on September 23, 1985. It was reviewed and updated again in September 1986 (see Appendix).

The introduction to the guidance document emphasizes the distinction between somatic cell gene therapy and germ line gene therapy and states that the RAC and its subcommittee "will not at present entertain proposals for germ line alterations but will consider for approval protocols involving somatic cell gene therapy." LeRoy Walters, a member of the gene therapy subcommittee and director of the Center for Bioethics at Georgetown University, notes that the phrase "at present" reflects an appreciation of the fact that it is very difficult to predict the future of scientific development. Such a phrase, he adds, indicates an openness to new ideas, but also conveys the message that the burden of proof for establishing the desirability of new approaches to the treatment of human diseases rests with those who propose them.

The remainder of the "Points to Consider" document is divided into four parts. The first part deals with the short-term risks and benefits of the proposed research to the patient and to other people, as well as with issues of fairness in the selection of patients, informed consent, and privacy and confidentiality. The second part addresses special issues related to the free flow of information regarding gene therapy trials. (Researchers are asked what steps they will take, consistent with the need to protect the privacy of gene therapy patients, to ensure that accurate information is made available to the public.) The third part summarizes other requested documentation, and the fourth part specifies reporting requirements. Researchers will be required to report twice yearly on the general progress of gene therapy patients (for at least three to five years) to the local IRB and to the NIH Office of Recombinant DNA Activities.

Review at the National Level

The review process established by RAC (see Figure 8.1) is designed to ensure that all interested persons have access to information about early gene therapy proposals and the manner in which they are evaluated. Each proposal will be summarized in lay language in the *Federal Register* prior to consideration by the Human Gene Therapy Subcommittee—this requirement is unprecedented in the regulation of medical technologies. The full RAC will review both the subcommittee's report and public comments before forwarding its recommendations to the director of NIH, who will make the final decision to approve or reject each proposal within the framework established by the "Points to Consider" document.

U.S. Food and Drug Administration

The public review by RAC will complement the confidential review of gene therapy proposals conducted by the U.S. Food and Drug Administration under its mandate to regulate new biological drugs.[4] The nucleic acids (DNA or RNA) and viruses employed in human gene therapy will be subject to the same requirements as other biological drugs intended for human use.

An FDA policy statement published in the *Federal Register* in June 1986 explains that investigators planning to use new drugs must file a Notice of Claimed Investigational Exemption for a New Drug (IND) to conduct clinical investigations on human subjects. The IND must contain information on drug composition, manufacturing and controls data, results of animal testing, training and experience of investigators, and a plan for clinical investigation. Researchers also must detail procedures for obtaining informed consent and for protecting the rights and safety of their subjects.

During the course of a clinical trial, investigators must furnish progress reports to the FDA at least once a year. Any adverse effects of the new therapy must be reported promptly.[5]

APPROVAL PROCESS FOR HUMAN GENE THERAPY STUDIES

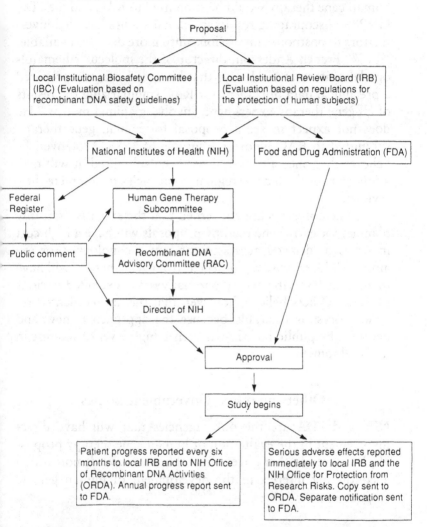

FIGURE 8.1 The NIH and FDA review process for clinical studies involving human somatic cell gene therapy.

Timetable

Although many scientists expected that the first proposals for human gene therapy would be submitted to RAC and the FDA in 1986, discouraging results in animal studies have led investigators to postpone submissions until more data are available.

W. French Anderson, director of the molecular hematology laboratory at NIH, says that his group recently prepared a preclinical data document for RAC summarizing the results of all gene therapy experiments involving animal models.[6] He does not expect to see a proposal for human gene therapy submitted to RAC before the middle of 1988. Moreover, he believes that once a proposal has been submitted, it will take a full year for the federal regulatory agencies to complete their reviews.

Some observers are concerned that the extensive scrutiny planned for early gene therapy proposals will have a high cost in terms of delayed access to the new technology. (For example, children who might benefit from gene therapy will have to rely on less satisfactory alternatives for extended periods of time.) Others believe, however, that the unprecedented review process is desirable because the approach is new and because the public has shown such a high level of interest in its development.

Other Interested Government Bodies

NIH and FDA are the only agencies that will have direct involvement in the evaluation of human gene therapy proposals, but several other governmental bodies have demonstrated a continuing interest in the development of human genetic engineering.

Domestic Policy Council Working Group on Biotechnology

In 1984, the Cabinet Council on Natural Resources and the Environment formed a special working group to determine whether existing regulations and policies were sufficient to ensure the safety of biotechnology research and products. The working group, originally called the Cabinet Council Working Group on Biotechnology (now the Domestic Policy Council Working Group on Biotechnology), was placed under the leadership of the White House Office of Science and Technology Policy.

Following an eight-month study, the Working Group published a detailed report with two major conclusions. First, existing legislation provided federal agencies (including FDA, the U.S. Department of Agriculture [USDA], the Environmental Protection Agency [EPA], and the Occupational Safety and Health Administration [OSHA]) with adequate regulatory authority to deal with the emerging processes and products of modern biotechnology. Second, an interagency coordinating committee was needed to encourage sharing of scientific information among the agencies, and to foster uniform standards of scientific review. (See description of the Biotechnology Science Coordinating Committee, below.)

The Domestic Policy Council Working Group on Biotechnology will continue to monitor developments in all areas of biotechnology, including human gene therapy. For now, the consensus is clearly that NIH and FDA oversight of gene therapy is sufficient, and that no new administrative or legislative actions are required.

Biotechnology Science Coordinating Committee

The Biotechnology Science Coordinating Committee (BSCC), established October 31, 1985, consists of senior policy officials from the National Science Foundation, EPA, FDA, NIH, and USDA. As part of the Federal Coordinating Council for Science, Engineering and Technology, it is managed by the Office of Science and Technology Policy, Executive Office of the President.

The BSCC is unlikely to take any actions on gene therapy in the near future, unless differences arise between NIH and FDA over the evaluation of proposals for clinical trials. In that case, the BSCC would provide a forum for resolving those differences.

Congressional Activities Related to Gene Therapy

Congressional interest in the social and ethical ramifications of human gene therapy began in the late 1970s and developed into a major focus of concern during the early 1980s. Following the publication of *Splicing Life* in November 1982, Representative Albert Gore, Jr., presided over three days of hearings on "Human Genetic Engineering." In 1984, the Subcommittee on Investigations and Oversight of the House Committee on Science and Technology, chaired by Mr. Gore, asked the Office of Technology Assessment to conduct a study of gene therapy, which resulted in the comprehensive report *Human Gene Therapy: A Background Paper.*[7]

In both 1984 and 1985, NIH authorization bills included a proposal for a new Biomedical Ethics Board to consist of six members of the Senate and six members of the House of Representatives. President Reagan vetoed both bills for reasons unrelated to the Ethics Board, but Congress overrode the veto in 1985.

Biomedical Ethics Board. The Biomedical Ethics Board is required by law to study and report to the Congress on ethical issues arising from the delivery of health care and biomedical and behavioral research. To conduct the studies and make the reports, it is empowered to appoint a Biomedical Ethics Advisory Committee consisting of fourteen members: four from the fields of biomedical or behavioral research; three physicians or others involved in the provision of health care; five distinguished in one or more of the fields of ethics, theology, law, the natural sciences, the social sciences, the humanities, health administration, government, or public affairs; and two lay persons representative of citizens with an interest in biomedical ethics.

In general, the Board has great latitude in identifying appropriate subjects for study. However, within 18 months of the appointment of the Advisory Committee, it must "deliver a report on research and developments in genetic engineering (including activities in recombinant DNA technology) which have implications for human genetic engineering."[8]

The process of selecting members for the Advisory Committee has extended beyond one term of Congress, but is close to completion. The Board's report on gene therapy and other aspects of human genetic engineering may be available by early 1989.

International Activities on Human Gene Therapy

The public policy debate about human gene therapy has not been limited to the United States.[9] In 1980, several members of the Council of Europe's Parliamentary Assembly called for action on the dual questions of genetic engineering and artificial reproduction. Their motion was referred to the Assembly's Legal Affairs and Science and Technology committees, which reported back to the Parliamentary Assembly in January 1982.

The Assembly focused primarily on germ line gene therapy. It recommended that its Committee of Ministers undertake three tasks:

(1) Draw up a European agreement on what constitutes legitimate application to human beings (including future generations) of the techniques of genetic engineering . . .

(2) Provide for explicit recognition of the right to a genetic inheritance which has not been artificially interfered with, except in accordance with certain principles which are recognized as being fully compatible with respect for human rights (as, for example, in the field of therapeutic applications), [and]

(3) Provide for the drawing up of a list of serious diseases which may properly, with the consent of the person concerned, be treated by gene therapy (though certain uses without consent, in line with existing practice for other forms of medical treatment, may be recognised as compatible with respect for human rights in the probability of a very serious disease being transmitted to a person's offspring).[10]

LeRoy Walters explains that the Assembly asserted the right of every individual to inherit a genetic pattern that has not been artificially changed, but "at the same time accepted germ-line intervention for the treatment and prevention of a canon-like list of serious diseases." The expert committee appointed by the Ministers to study how the Assembly's recommendations could be implemented has not yet issued a formal report.

Since the Assembly's deliberations, national committees in Sweden, the Federal Republic of Germany, the United Kingdom, and France independently have addressed issues arising from human genetic engineering. The Swedish and West German committees both concluded that the use of human somatic cell gene therapy to treat genetic diseases would be acceptable (and no different, in ethical terms, from organ transplantation and other existing forms of therapy).

The four committees had somewhat different responses to germ line gene therapy. The Swedish committee said that if gene therapy on human sperm, ova, and early embryos became possible to perform in a reliable fashion, "then the operation must come under a severe ethical examination which should include full knowledge of all the consequences."[11] The Committee of Inquiry into Human Fertilisation and Embryology in the United Kingdom recommended that a new governmental licensing agency (created primarily to oversee clinical in vitro fertilization and research on early human embryos) consider the issues raised by the genetic alteration of embryos.[12]

The French National Ethics Committee and the West German commission both concluded that research involving human germ line gene therapy should be prohibited.[13] However, the commission added, "It cannot be denied that possibly in the future developments may occur that could make it necessary to loosen [this] general prohibition in the interest of protecting life and health."[14]

Standard Medical Practice

This chapter has focused primarily on the regulation of early stages in the application of genetic engineering to human beings. LeRoy Walters predicts that if somatic cell gene therapy proves to be both safe and effective, policymakers will opt for gradual decentralization of the review process for somatic cell proposals. As local IRBs and IBCs gain more confidence in dealing with gene therapy (and public understanding of the benefits and risks increases), the need for case-by-case review by RAC and the NIH director may gradually diminish.[15] (A similar evolution has occurred in the regulation of basic science experiments involving recombinant DNA. Today, only a very small fraction of these experiments are subject to RAC review.)

Authors of the OTA report, *Human Gene Therapy*, note that if gene therapy becomes a part of standard medical practice, it will be subject to the same regulatory concerns common to other types of routine therapies. Among the questions that will arise are

- Who will pay for gene therapy?
- How will the health care system provide equitable access to gene therapy?
- Who will be considered qualified to perform the procedure?
- Who will be responsible for ensuring that its use falls within accepted social guidelines?

One determinant of the future distribution of benefits from gene therapy will be the way in which gene therapy research develops over the next decade. Many genetic diseases are more prevalent in some populations (that is, racial, religious, or ethnic groups) than in others. Current research efforts in gene therapy focus primarily on the search for general solutions to complex technical problems. But as scientists learn more about vector design and gene expression, gene therapy research could become increasingly disease specific. In the future, federal decisions about which disease-specific projects to fund could play a role in determining who eventually will benefit from the new technology.

Conclusions

The policy issues raised by human gene therapy have received extensive national and international attention. The consensus among policymakers in the United States and elsewhere is that

somatic cell gene therapy does not create any new social or
ethical dilemmas. Nonetheless, widespread public interest in
the applications of the new genetic technologies to human
beings has prompted the development in the United States of
a special public review process for early clinical trials of gene
therapy. This process will complement existing measures that
safeguard the rights of all patients involved in human experimentation.

Five years ago, the President's Commission for the Study
of Ethical Problems in Medicine and Biomedical and Behavioral Research expressed concern that no body within the
federal government was prepared to deal with the complex
range of issues raised by the projected human uses of gene
splicing. This is no longer the case. Both the executive branch
and the legislative branch have established formats for assessing the possible long-term consequences of current research
efforts in human genetic engineering.

—— ACKNOWLEDGMENTS ——

Chapter 8 is based on the presentation of LeRoy Walters.

—— NOTES ——

1. The first nonscientists appointed to RAC were Emmette Red-
ford, a professor of government at the University of Texas at Austin,
and LeRoy Walters, director of the Center for Bioethics at the
Kennedy Institute of Ethics, Georgetown University.

2. Elizabeth A. Milewski, "Development of a Points to Consider
Document for Human Somatic Cell Gene Therapy," *Recombinant
DNA Technical Bulletin* 8(4):176-180.

3. Milewski, p. 176.

4. The FDA mandate covers all investigators, whether or not
they are at an institution receiving federal funds for recombinant
DNA research.

5. Deaths or life-threatening events must be reported by telephone within three days. Less serious adverse effects must be reported by telephone or by letter within 10 working days.

6. The preclinical data document was prepared in response to
a memorandum, dated February 25, 1986, in which the Executive
Secretary of the Human Gene Therapy Subcommitte notified re-

searchers that the subcommittee was interested in preclinical data from studies leading toward the development of techniques for human somatic cell gene therapy. The memorandum emphasized that the submission of such data would not be a part of the formal review and approval process, but rather would be part of an educational process for subcommittee members and for the research community.

7. As noted in Chapter 1, the OTA report concluded, "Civic, religious, scientific, and medical groups have all accepted, in principle, the appropriateness of gene therapy of somatic cells in humans for specific genetic diseases. Somatic cell gene therapy is seen as an extension of present methods of therapy that might be preferable to other technologies" (*Human Gene Therapy*, p. 47).

8. The only other required study is a review of federal policies on fetal research.

9. In fact, LeRoy Walters says, "It is possible that more ethical and public policy analysis has occurred before the fact with human gene therapy than with any other biomedical technology in history."

10. Parliamentary Assembly of the Council of Europe, Recommendation 934 (1982) on Genetic Engineering, *Texts Adopted by the Assembly* (Strasbourg: Council of Europe).

11. Sweden, Gen-etikkommitten, *Genetisk Integritet*, p. 23, Statens Offentliga Utredningar, Socialdepartementet (Stockholm: Liber Tryck, 1984).

12. *Human Gene Therapy*, p. 35.

13. P. Lewis, "Ban Urged for Some Gene Research," *New York Times*, December 16, 1986, p. C16.

14. Federal Republic of Germany, Bundesminister der Justiz und Bundesminister für Forschung und Technologie, *In-vitro-Fertilisation, Genomanalyse und Gentherapie: Bericht der Arbeitsgruppe*, translated by LeRoy Walters (Bonn: Bundesminister der Justiz und Bundesminister für Forschung und Technologie, 1985).

15. The FDA review process is mandated by law and would not change.

—— SUGGESTED READING ——

Culliton, Barbara J. 1986. "NIH Asked to Tighten Gene Therapy Rules." *Science* 233:1378-1379.

Federal Republic of Germany, Bundesminister der Justiz und Bundesminister für Forschung und Technologie. 1985. *In-vitro-Fertilisation, Genomanalyse und Gentherapie: Bericht der Arbeitsgruppe*. Bonn: Bundesminister der Justiz und Bundesminister für Forschung und Technologie.

Food and Drug Administration. 1986. "Statement of Policy for Regulating Biotechnology Products." *Federal Register* 51(123):23309-23311.

Health Research Extension Act of 1985. Public Law 99-158. November 20, 1985. (42 USC 275) Section 381. Biomedical Ethics.

Human Gene Therapy—A Background Paper. 1984. Washington, D.C.: U.S. Congress, Office of Technology Assessment. OTA-BP-BA-32.

Krimsky, Sheldon. 1982. *Genetic Alchemy: The Social History of the Recombinant DNA Controversy*. Cambridge, Massachusetts: MIT Press.

Lewis, Paul. 1986. "Ban Urged for Some Gene Research." *New York Times*. December 16, p. C16.

Milewski, Elizabeth A. 1985. "Development of a Points to Consider Document for Human Somatic Cell Gene Therapy." *Recombinant DNA Technical Bulletin* 8(4):176-180.

Miller, Henry I. 1984. "Gene Therapy: Overregulation May Be Hazardous to Our Health." *Gene* 27:1-2.

NIH Recombinant DNA Advisory Committee. 1986. "Points to Consider in the Design and Submission of Human Somatic-Cell Gene Therapy Proposals." Adopted September 29. (See Appendix.)

Office of Science and Technology Policy. 1986. "Coordinated Framework for Regulation of Biotechnology." *Federal Register* 51(123):23302-23309.

Olson, Steve. 1986. *Biotechnology: An Industry Comes of Age*. Washington, D.C.: National Academy Press.

Parliamentary Assembly of the Council of Europe. 1982. "Recommendation 934 (1982) on Genetic Engineering." *Texts Adopted by the Assembly*. Strasbourg: Council of Europe.

President's Commission for the Study of Ethical Problems in Medicine and Biomedical and Behavioral Research. 1982. *Splicing Life: The Social and Ethical Issues of Genetic Engineering with Human Beings*. Washington, D.C.: U.S. Government Printing Office. Stock no. 83-600500.

Sweden, Gen-etikkommitten. 1984. *Genetisk Integritet*, p. 23. Statens Offentliga Utredningar, Socialdepartementet. Stockholm: Liber Tryck.

Walters, LeRoy. 1986. "The Ethics of Human Gene Therapy." *Nature* 320:225-227.

9 | Prospects for the Future

Modern medicine and surgery have relatively little to offer most children born with severe genetic disorders. Dietary and pharmacologic therapies and new techniques in transplant surgery allow a small minority to live normal or near-normal lives, but the majority face repeated hospitalizations and, often, an early death. Children with diseases such as sickle cell anemia, cystic fibrosis, thalassemia, and severe combined immune deficiency live longer than they did in the past, but their lives remain bound by restrictions imposed by their illnesses.

Somatic cell gene therapy is one of several new therapeutic approaches that may help such children in the future. A child with a life-threatening but reversible genetic disease caused by a defect in a single gene would be treated with the gene's normal counterpart. The normal gene, provided through recombinant DNA technology, would be inserted into a specific tissue in the child's body and would not be passed on to future generations.

Ethical and Social Concerns

The potential application of genetic engineering to human beings has generated widespread public attention. Some segments of society have expressed concern about the ethical and social implications of manipulating human genes. During a decade of public discussion, experts in biomedical ethics have developed a framework to help put these concerns in perspective. The framework divides gene therapy into four distinct categories: somatic cell gene therapy, germ line gene therapy, enhancement genetic engineering, and eugenic genetic engineering.

Somatic cell gene therapy has the same goals as all other forms of medical treatment: restoration of health and reduction of human suffering. In practice, it will be very similar to present-day bone marrow and organ transplantation.

The ethical issues raised by somatic cell gene therapy are identical to those raised by other novel therapeutic approaches. Prior to clinical trials, researchers will have to demonstrate that the potential benefits of the procedure are greater than the potential risks; that therapies known to be effective will not be withheld for purposes of conducting the trials; that the process of selecting patients will be fair; and that prospective subjects and their parents will be fully informed about potential adverse effects. Once these conditions have been met, there will no reason to refrain from human studies of the new technique.

In contrast, germ line gene therapy, enhancement genetic engineering, and eugenic genetic engineering raise scientific and ethical issues beyond those associated with other medical technologies. For example, in germ line gene therapy, genetic changes would be passed on to future generations. Most scientists and policymakers believe that germ line gene therapy should not be undertaken without additional public discussion, because its effects would extend beyond the individual patient to those who could not possibly give informed consent.

Enhancement genetic engineering refers to the use of gene transfer to change a specific characteristic, such as height, in a healthy child. The prospect of parents turning to enhancement techniques to produce "perfect" children raises important questions about the value of the individual in our society and about the appropriate use of limited health care resources. Thus, society as a whole must decide whether enhancement of physical traits is an acceptable rationale for medically manipulating human genes.

The underlying fear among those who object most strenuously to the use of gene therapy is that genetic engineering eventually might be employed by an amoral government to alter complex traits such as intelligence and personality (which involve hundreds, or perhaps thousands, of genes interacting with multiple factors in the environment). In fact, such eugenic genetic engineering would require tools far more sophisticated than those available to today's scientists.

Discussions of eugenic genetic engineering continue to be useful to help clarify concerns about the impact of future technologies on human society, but such discussions should not be viewed as an indication that eugenic genetic engineering is either imminent or a goal of current research. Indeed, eugenic genetic engineering using recombinant DNA techniques may never be possible because of the extreme complexity of the systems involved. Diverse governmental, civic, religious, scientific, and medical bodies in the United States and Europe have concluded that it would be unethical to withhold somatic cell gene therapy from severely ill patients solely because other forms of genetic engineering might be misused in the future.

Early Applications

The first human studies of somatic cell gene therapy probably will involve the insertion of normal genes into bone marrow cells, using a modified virus derived from the family known as retroviruses. Many researchers believe that two inherited disorders of the immune system, adenosine deaminase (ADA) deficiency and purine nucleoside phosphorylase (PNP) deficiency, are the most likely disease candidates for these early clinical trials. Both diseases appear to meet basic criteria for gene therapy studies: they are fatal in early childhood; the cell types affected by the disorders have been identified; and the normal counterparts of the defective genes have been isolated and copied in the laboratory.

The choice of disease candidates and potential target cells for the first attempts at somatic cell gene therapy could change, however, as a result of new developments in genetic engineering and other fields. For example, a new form of enzyme replacement therapy recently has led to improved immune function in several children with ADA deficiency. If enzyme replacement proves to be a reasonable long-term solution for these patients, the disease may no longer be a suitable candidate for early gene therapy trials.

Potential Target Cells

Experiments in several laboratories also have raised questions about the dominant role of bone marrow studies in gene therapy research. Bone marrow cells go through many different stages of development. At each stage, some genes are activated and others are switched off to allow cells to meet new developmental demands. Recent evidence suggests that this maturation process may interfere with the expression of foreign genes.

Other important observations concern the process of blood formation. Blood cells are derived from immature bone marrow cells called stem cells. Studies in mice indicate that the majority of blood cells present in an animal after bone marrow transplantation are derived from just a few stem cells, although a bone marrow graft probably contains hundreds of such cells. In addition, the contribution of individual stem cells to blood formation appears to change over time. Together, these findings suggest that it might be difficult to achieve consistent therapeutic responses in patients treated with gene transfer into bone marrow cells.

These results do not mean that bone marrow will not be a suitable target tissue for gene therapy or that research on bone marrow cells should stop. Instead, they underscore the need for more basic information about the factors that control blood cell development in laboratory animals and in human beings. They also highlight the need for continued investigation of other cell types that might be suitable targets for gene therapy, such as skin cells and liver cells.

Scientists also stress the need for more information about natural and modified retroviruses and other potential mechanisms for inserting genes into cells. Researchers have just begun to explore the range of possible structures for retroviral vectors. Many more must be tested to understand why some succeed and others fail. Two important messages can be derived from existing studies: (1) researchers cannot assume that a retroviral vector that results in the expression of one human gene will function the same way with a different gene, and (2) vectors that work in one animal system may not work in another.

Animal Models

Some experimental problems may be overcome by the development of better animal models of the relevant genetic diseases. Laboratories worldwide are working to identify natural animal models of diseases such as ADA deficiency and sickle cell anemia. In addition, advances in genetic engineering will make it possible to produce new animal models by disabling specific genes in target cells.

The consensus among scientists in the field of gene therapy is that recent studies have provided valuable new information on which to base future research efforts, but they have not satisfied important prerequisites for human studies. Gene therapy will not become a reasonable alternative for patient care until researchers can achieve higher and more stable levels of expression of inserted genes in laboratory animals.

Review Policy

Predictions about when these goals will be met range from two years to more than ten years. Local and national review boards will decide when research in gene therapy has progressed sufficiently to justify clinical trials. These boards consist of laboratory scientists, practicing physicians, bioethicists, attorneys, specialists in public policy, and representatives of the lay public. Their broad membership will ensure a comprehensive approach to the full range of scientific, ethical, and social issues raised by the introduction of genetic engineering into medical practice.

The early use of somatic cell gene therapy will be limited to a small number of inherited metabolic disorders, but its ultimate contribution to human health could be much greater.

Appendix

NATIONAL INSTITUTES OF HEALTH
POINTS TO CONSIDER IN THE DESIGN AND SUBMISSION OF
HUMAN SOMATIC-CELL GENE THERAPY PROTOCOLS

HUMAN GENE THERAPY SUBCOMMITTEE
NIH RECOMBINANT DNA ADVISORY COMMITTEE

Applicability - These "Points to Consider" apply only to research
conducted at or sponsored by an institution that receives any support for recombinant DNA research from the National Institutes of Health (NIH). This includes research performed by NIH directly.

Introduction

(1) Experiments in which recombinant DNA[1] is introduced into
cells of a human subject with the intent of stably modifying the subject's genome are covered by Section III-A-4 of the NIH Guidelines for Research Involving Recombinant DNA Molecules (49 Federal Register 46266). Section III-A-4 requires such experiments to be reviewed by the NIH Recombinant DNA Advisory Committee (RAC) and approved by the NIH. RAC consideration of each proposal will be on a case-by-case basis and will follow publication of a precis of the proposal in the *Federal Register*, an opportunity for public comment, and a review of the proposal by the working group of the RAC. RAC recommendations on each proposal will be forwarded to the

---

[1] Section III-A-4 applies both to recombinant DNA and to DNA or RNA derived from recombinant DNA.

NIH Director for a decision which will then be published in the *Federal Register*. In accordance with Section IV-C-1-b of the NIH Guidelines, the NIH Director may approve proposals only if he finds that they present "no significant risk to health or the environment."

(2) In general, it is expected that somatic-cell gene therapy protocols will not present a risk to the environment as the recombinant DNA is expected to be confined to the human subject. Nevertheless, Section I-B-4-b of the "Points to Consider" document asks the researchers to address specifically this point.

(3) This document is intended to provide guidance in preparing proposals for NIH consideration under Section III-A-4 of the NIH Guidelines for Research Involving Recombinant DNA Molecules. Not every point mentioned in the "Points to Consider" document will necessarily require attention in every proposal. The document will be considered for revision as experience in evaluating proposals accumulates and as new scientific developments occur. This review will be carried out at least annually.

(4) A proposal will be considered by the RAC only after the protocol has been approved by the local Institutional Biosafety Committee (IBC) and by the local Institutional Review Board (IRB) in accordance with Department of Health and Human Services (DHHS) Regulations for the Protection of Human Subjects (45 Code of Federal Regulations, Part 46). If a proposal involves children, special attention should be paid to subpart D of these DHHS regulations. The IRB and IBC may, at their discretion, condition their approval on further specific deliberation by the RAC and its working group. Consideration of gene therapy proposals by the RAC may proceed simultaneously with review by any other involved federal agencies[2] provided that the RAC is notified of the simultaneous review. Meetings of the committee will be open to the public except where trade secrets or proprietary information would be disclosed. The committee would prefer that the first proposals submitted for RAC review contain no proprietary information

---

[2] The Food and Drug Administration (FDA) has jurisdiction over drug products intended for use in clinical trials of human somatic-cell gene therapy. For general information on FDA's policies and regulatory requirements, please see the *Federal Register*, Volume 51, pages 23309–23313, 1986.

or trade secrets, enabling all aspects of the review to be open to the public. The public review of these protocols will serve to inform the public not only on the technical aspects of the proposals but also on the meaning and significance of the research.

(5) The clinical application of recombinant DNA techniques to human gene therapy raises two general kinds of questions: (1) the questions usually discussed by IRBs in their review of *any* proposed research involving human subjects; and (2) broader social issues. The first type of question is addressed principally in Part I of this document. Several of the broader social issues surrounding human gene therapy are discussed later in this Introduction and in Part II below.

(6) Following the Introduction, this document is divided into four parts. Part I deals with the short-term risks and benefits of the proposed research to the patient[3] and to other people, as well as with issues of fairness in the selection of patients, informed consent, and privacy and confidentiality. In Part II, investigators are requested to address special issues pertaining to the free flow of information about clinical trials of gene therapy. These issues lie outside the usual purview of IRBs and reflect general public concerns about biomedical research. Part III summarizes other requested documentation that will assist the RAC and its working group in their review of gene therapy proposals. Part IV specifies reporting requirements.

(7) A distinction should be drawn between making genetic changes in somatic cells and in germ line cells. The purpose of somatic cell gene therapy is to treat an individual patient, e.g., by inserting a properly functioning gene into a patient's bone marrow cells *in vitro* and then reintroducing the cells into the patient's body. In germ line alterations, a specific attempt is made to introduce genetic changes into the germ (reproductive) cells of an individual, with the aim of changing the set of genes passed on to the individual's offspring. The RAC and its working group will not at present entertain proposals for germ line alterations but will consider for approval protocols involving somatic-cell gene therapy.

---

[3] The term "patient" and its variants are used in the text as a shorthand designation for "patient-subject."

(8) The acceptability of human somatic-cell gene therapy has been addressed in several recent public documents as well as in numerous academic studies. The November 1982 report of the President's Commission for the Study of Ethical Problems in Medicine and Biomedical and Behavioral Research, *Splicing Life*, resulted from a two-year process of public deliberations and hearings; upon release of that report, a House subcommittee held three days of public hearings with witnesses from a wide range of fields from the biomedical and social sciences to theology, philosophy, and law. In December 1984, the Office of Technology Assessment released a background paper, *Human Gene Therapy*, which brought these earlier documents up-to-date. As the latter report concluded:

> Civic, religious, scientific, and medical groups have all accepted, in principle, the appropriateness of gene therapy of somatic cells in humans for specific genetic diseases. Somatic cell gene therapy is seen as an extension of present methods of therapy that might be preferable to other technologies.

(9) Concurring with this judgment, the RAC and its working group are prepared to consider for approval somatic-cell therapy protocols, provided that the design of such experiments offers adequate assurance that their *consequences* will not go beyond their *purpose*, which is the same as the traditional purpose of all clinical investigations, namely, to benefit the health and well-being of the individual being treated while at the same time gathering generalizable knowledge.

(10) Two possible undesirable consequences of somatic-cell therapy would be unintentional (1) vertical transmission of genetic changes from an individual to his or her offspring or (2) horizontal transmission of viral infection to other persons with whom the individual comes in contact. Accordingly, this document requests information that will enable the RAC and its working group to assess the likelihood that the proposed somatic-cell gene therapy will inadvertently affect reproductive cells or lead to infection of other people (e.g., treatment personnel or relatives).

(11) In recognition of the social concern that surrounds the general discussion of human gene therapy, the working group will continue to consider the possible long-range effects of applying knowledge gained from these and related experiments. While research in molecular biology could lead to the development

of techniques for germ line intervention or for the use of genetic means to enhance human capabilities rather than to correct defects in patients, the working group does not believe that these effects will follow immediately or inevitably from experiments with somatic-cell gene therapy. The working group will cooperate with other groups in assessing the possible long-term consequences of somatic-cell gene therapy and related laboratory and animal experiments in order to define appropriate human applications of this emerging technology.

(12) Responses to the questions raised in these "Points to Consider" should be provided in the form of either written answers or references to specific sections of the protocol or its appendices.

I. Description of Proposal

A. Objectives and rationale of the proposed research

State concisely the overall objectives and rationale of the proposed study. Please provide information on the following specific points:

1. Why is the disease selected for treatment by means of gene therapy a good candidate for such treatment?

2. Describe the natural history and range of expression of the disease selected for treatment. What objective and/or quantitative measures of disease activity are available? In your view, are the usual effects of the disease predictable enough to allow for meaningful assessment of the results of gene therapy?

3. Is the protocol designed to prevent all manifestations of the disease, to halt the progression of the disease after symptoms have begun to appear, or to reverse manifestations of the disease in seriously ill victims?

4. What alternative therapies exist? In what groups of patients are these therapies effective? What are their relative advantages and disadvantages as compared with the proposed gene therapy?

B. Research design, anticipated risks and benefits

1. *Structure and characteristics of the biological system*
   Provide a full description of the methods and reagents to

be employed for gene delivery and the rationale for their use. The following are specific points to be addressed:

a. What is the structure of the cloned DNA that will be used?

   (1) Describe the gene (genomic or cDNA), the bacterial plasmid or phage vector, and the delivery vector (if any). Provide complete nucleotide sequence analysis or a detailed restriction enzyme map of the total construct.

   (2) What regulatory elements does the construct contain (e.g., promoters, enhancers, polyadenylation sites, replication origins, etc.)?

   (3) Describe the steps used to derive the DNA construct.

b. What is the structure of the material that will be administered to the patient?

   (1) Describe the preparation, structure, and composition of the materials that will be given to the patient or used to treat the patient's cells.

      (a) If DNA, what is the purity (both in terms of being a single DNA species and in terms of other contaminants)? What tests have been used and what is the sensitivity of the tests?

      (b) If a virus, how is it prepared from the DNA construct? In what cell is the virus grown (any special features)? What medium and serum are used? How is the virus purified? What is its structure and purity? What steps are being taken (and assays used with their sensitivity) to detect and eliminate any contaminating materials (for example, VL30 RNA, other nucleic acids, or proteins) or contaminating viruses or other organisms in the cells or serum used for preparation of the virus stock?

      (c) If co-cultivation is employed, what kinds of cells are being used for co-cultivation? What steps are being taken (and assays used with their sensitivity) to detect and eliminate any contaminating materials? Specifically, what tests are being done to assess the material to be returned to the patient for the presence of live or killed donor cells or other non-vector materials (for example, VL30 sequences) originating from those cells?

(d) If methods other than those covered by (a)–(c) are used to introduce new genetic information into target cells, what steps are being taken to detect and eliminate any contaminating materials? What are possible sources of contamination? What is the sensitivity of tests used to monitor contamination?

(2) Describe any other material to be used in preparation of the material to be administered to the patient. For example, if a viral vector is proposed, what is the nature of the helper virus or cell line? If carrier particles are to be used, what is the nature of these?

2. *Preclinical studies, including risk-assessment studies*
Describe the experimental basis (derived from tests in cultured cells and animals) for claims about the efficacy and safety of the proposed system for gene delivery.

a. *Laboratory studies of the delivery system*

(1) What cells are the intended recipients of gene therapy? If recipient cells are to be treated *in vitro* and returned to the patient, how will the cells be characterized before and after treatment? What is the theoretical and practical basis for assuming that only the treated cells will act as recipients?

(2) Is the delivery system efficient? What percentage of the target cells contain the added DNA?

(3) How is the structure of the added DNA sequences monitored and what is the sensitivity of the analysis? Is the added DNA extrachromosomal or integrated? Is the added DNA unrearranged?

(4) How many copies are present per cell? How stable is the added DNA both in terms of its continued presence and its structural stability?

b. *Laboratory studies of gene expression*

Is the added gene expressed? To what extent is expression only from the desired gene (and not from the surrounding DNA)? In what percentage of cells does expression from the added DNA occur? Is the product biologically active? What percentage of normal activity results from the inserted gene? Is the gene expressed in cells other than the target cells? If so, to what extent?

c. *Laboratory studies pertaining to the safety of the delivery/expression system*

(1) If a retroviral system is used:

(a) What cell types have been infected with the retroviral vector preparation? Which cells, if any, produce infectious particles?

(b) How stable are the retroviral vector and the resulting provirus against loss, rearrangement, recombination, or mutation? What information is available on how much rearrangement or recombination with endogenous or other viral sequences is likely to occur in the patient's cells? What steps have been taken in designing the vector to minimize instability or variation? What laboratory studies have been performed to check for stability, and what is the sensitivity of the analyses?

(c) What laboratory evidence is available concerning potential harmful effects of the treatment, e.g., development of neoplasia, harmful mutations, regeneration of infectious particles, or immune responses? What steps have been taken in designing the vector to minimize pathogenicity? What laboratory studies have been performed to check for pathogenicity, and what is the sensitivity of the analyses?

(d) Is there evidence from animal studies that vector DNA has entered untreated cells, particularly germ line cells? What is the sensitivity of the analyses?

(e) Has a protocol similar to the one proposed for a clinical trial been carried out in non-human primates and/or other animals? What were the results? Specifically, is there any evidence that the retroviral vector has recombined with any endogenous or other viral sequences in the animals?

(2) If a non-retroviral delivery system is used: What animal studies have been done to determine if there are pathological or other undesirable consequences of the protocol (including insertion of DNA into cells other than those treated, particularly germ line cells)? How long have the animals been studied after treatment? What tests have been used and what is their sensitivity?

3. *Clinical procedures, including patient monitoring*
Describe the treatment that will be administered to patients and the diagnostic methods that will be used to monitor the success or failure of the treatment. If previous clinical studies using similar methods have been performed by yourself or others, indicate their relevance to the proposed study.

a. Will cells (e.g., bone marrow cells) be removed from patients and treated *in vitro* in preparation for gene therapy? If so, what kinds of cells will be removed from the patients, how many, how often, and at what intervals?

b. Will patients be treated to eliminate or reduce the number of cells containing malfunctioning genes (e.g., through radiation or chemotherapy) prior to gene therapy?

c. What treated cells (or vector/DNA combination) will be given to patients in the attempt to administer gene therapy? How will the treated cells be administered?

What volume of cells will be used? Will there be single or multiple treatments? If so, over what period of time?

d. What are the clinical endpoints of the study? Are there objective and quantitative measurements to assess the natural history of the disease? Will such measurements be used in following your patients? How will patients be monitored to assess specific effects of the treatment on the disease? What is the sensitivity of the analyses? How frequently will follow-up studies be done? How long will patient follow-up continue?

e. What are the major potential beneficial and adverse effects of treatment that you anticipate? What measures will be taken in an attempt to control or reverse these adverse effects if they occur? Compare the probability and magnitude of potential adverse effects on patients with the probability and magnitude of deleterious consequences from the disease if gene therapy is not performed.

f. If a treated patient dies, what special studies will be performed as part of the autopsy?

4. *Public-health considerations*
Describe any potential benefits and hazards of the proposed therapy to persons other than the patients being treated. Specifically:

a. On what basis are potential public health benefits or hazards postulated?

b. Is there a significant likelihood that the added DNA will spread from the patient to other persons or to the environment?

c. What precautions will be taken against such spread (e.g., to patients sharing a room, health-care workers, or family members)?

d. What measures will be undertaken to mitigate the risks, if any, to public health?

5. *Qualifications of investigators, adequacy of laboratory and clinical facilities*

Indicate the relevant training and experience of the personnel who will be involved in the preclinical studies and clinical administration of gene therapy. In addition, please describe the laboratory and clinical facilities where the proposed study will be performed.

a. What professional personnel (medical and nonmedical) will be involved in the proposed study? What are their specific qualifications and experience with respect to the disease to be treated and with respect to the techniques employed in molecular biology? Please provide *curricula vitae* (see Section III-E).

b. At what hospital or clinic will the treatment be given? Which facilities of the hospital or clinic will be especially important for the proposed study? Will patients occupy regular hospital beds or clinical research center beds? Where will patients reside during the follow-up period?

C. *Selection of patients*
Estimate the number of patients to be involved in the proposed study of gene therapy. Describe recruitment procedures and patient eligibility requirements, paying particular attention to whether these procedures and requirements are fair and equitable.

1. How many patients do you plan to involve in the proposed study?

2. How many eligible patients do you anticipate being able to identify each year?

3. What recruitment procedures do you plan to use?

4. What selection criteria do you plan to employ? What are the exclusion and inclusion criteria for the study?

5. How will patients be selected if it is not possible to include all who desire to participate?

D. *Informed consent*
Indicate how patients will be informed about the proposed study and how their consent will be solicited. The consent procedure should adhere to the requirements of DHHS regulations for the protection of human subjects (45 Code of

Federal Regulations, Part 46). If the study involves pediatric
or mentally handicapped patients, describe procedures for
seeking the permission of parents or guardians and, where
applicable, the assent of each patient. Areas of special concern highlighted below include potential adverse effects, financial costs, privacy, and long-term follow-up.

1. How will the major points covered in Sections I-A
   through I-C of this document be disclosed to potential
   participants in this study and/or parents or guardians in
   language that is understandable to them?

2. How will the innovative character and the theoretically-
   possible adverse effects of gene therapy be discussed with
   patients and/or parents or guardians? How will the potential adverse effects be compared with the consequences of the disease? What will be said to convey that
   some of these adverse effects, if they occur, could be
   irreversible?

3. What explanation of the financial costs of gene therapy
   and any available alternative therapies will be provided
   to patients and/or parents or guardians?

4. How will patients and/or their parents or guardians be
   informed that the innovative character of gene therapy
   may lead to great interest by the media in the research
   and in treated patients?

5. How will patients and/or their parents or guardians be
   informed:

   a. That some of the procedures performed in the study
      may be irreversible?

   b. That following the performance of such procedures it
      would not be medically advisable for patients to with-
      draw from the study?

   c. That a willingness to cooperate in long-term follow-
      up (for at least three to five years) will be a precon-
      dition for participation in the study?

   d. That a willingness to permit an autopsy to be per-
      formed in the event of a patient's death following
      treatment is also a precondition for a patient's partic-

ipation in the study? (This stipulation is included because an accurate determination of the precise cause of a patient's death would be of vital importance to all future gene therapy patients.)

E. *Privacy and confidentiality*
Indicate what measures will be taken to protect the privacy of gene therapy patients and their families as well as to maintain the confidentiality of research data.

1. What provisions will be made to honor the wishes of individual patients (and the parents or guardians of pediatric or mentally handicapped patients) as to whether, when, or how the identity of patients is publicly disclosed?

2. What provision will be made to maintain the confidentiality of research data, at least in cases where data could be linked to individual patients?

II. *Special Issues*
Although the following issues are beyond the normal purview of local IRBs, the RAC and its working group request that investigators respond to questions A and B below.

A. What steps will be taken, consistent with point I-E above, to ensure that accurate information is made available to the public with respect to such public concerns as may arise from the proposed study?

B. Do you or your funding sources intend to protect under patent or trade secret laws either the products or the procedures developed in the proposed study? If so, what steps will be taken to permit as full communication as possible among investigators and clinicians concerning research methods and results?

III. *Requested Documentation*
In addition to responses to the questions raised in these "Points to Consider," please submit the following materials:

A. Your protocol as approved by your local IRB and IBC. The consent form, which must have IRB approval, should be submitted to the NIH only on request.

B. Local IRB and IBC minutes and recommendations that pertain to your protocol.

C. A one-page scientific abstract of the gene therapy protocol.

D. A one-page description of the proposed experiment in nontechnical language.

E. *Curricula vitae* for professional personnel.

F. An indication of other federal agencies to which the protocol is being submitted for review.

G. Any other material which you believe will aid in the review.

IV. *Reporting Requirements*

A. Serious adverse effects of treatment should be reported immediately to both your local IRB and the NIH Office for Protection from Research Risks, and a written report should be filed with both groups. A copy of the report should also be forwarded to the NIH Office of Recombinant DNA Activities (ORDA).

B. Reports regarding the general progress of patients should be filed at six-month intervals with both your local IRB and ORDA. These twice-yearly reports should continue for a sufficient period of time to allow observation of all major effects (at least three to five years). In the event of a patient's death, the autopsy report should be submitted to the IRB and ORDA.

Resources

This list of resources for information on genetic diseases is divided into three sections. The first section describes directories of centers that provide genetic services in the United States and abroad. The second section contains directories of voluntary organizations concerned with the medical and psychological impacts of genetic disorders and birth defects. The final section lists organizations that promote communication on issues related to genetic services.

Individuals and families seeking specific information or referrals to health care providers in a particular area are encouraged to contact their local chapter of the March of Dimes Birth Defects Foundation (in the white pages of the telephone book). Also, the departments of public health in many states have genetics programs that provide information and referrals.

MEDICAL SERVICE DIRECTORIES

Comprehensive Clinical Genetic Services Centers: A National Directory. 1985. Washington, D.C.: National Center for Education in Maternal and Child Health.

This 107-page directory is "designed as a resource guide for administrators and health professionals who deal with individuals and families affected by or concerned with genetic disorders." It identifies the names, addresses, and contact persons of clinical genetic service centers throughout the United States that provide comprehensive diagnostic services, medical management, counseling, and follow-up care. Also listed are (1) state genetic service counselors, (2) state newborn screening directors, and (3) state directors of maternal and child health and crippled children's services.

The directory is available free of charge from the National

Center for Education in Maternal and Child Health, 38th and R Streets, N.W., Washington, D.C., 20057. (202) 625-8400.

International Directory of Genetic Services, Eighth Edition. 1986. Henry T. Lynch, William J. Kimberling, and Kathleen M. Brennan. White Plains, NY: March of Dimes Birth Defects Foundation.

This 57-page international directory cross-references medical genetic centers with the services they provide. The four major sections of the directory are: (1) directory of genetic units by country; (2) genetic units by director's last name; (3) genetic services by country; (4) availability of genetic services.

The directory is available for $2.00 from Professional Education, March of Dimes Birth Defects Foundation, 1275 Mamaroneck Avenue, White Plains, NY 10605. (914) 428-7100.

State Treatment Centers for Metabolic Disorders. 1986. Washington, D.C.: National Center for Education in Maternal and Child Health.

This 66-page directory lists clinical centers for metabolic disorders. The directory is divided into four sections: (1) state treatment centers for metabolic disorders; (2) state newborn screening directors; (3) state directors of maternal and child health and crippled children's services; and (4) coordinating and contributing regions for the maternal PKU collaborative study.

The directory is available free of charge from the National Center for Education in Maternal and Child Health, 38th and R Streets, NW, Washington, D.C. 20057. (202) 625-8400.

SUPPORT GROUPS

A Guide to Selected National Genetic Voluntary Organizations. 1986. Washington, D.C.: National Center for Education in Maternal and Child Health.

This 131-page directory provides a listing of "mutual support groups concerned with the medical and psychosocial impacts of genetic disorders and birth defects on affected individuals and families." The directory is arranged alphabetically by organizational name; it also includes a user's guide that is alphabetical by disorder name.

The directory is available for $5.00 from the National Center for Education in Maternal and Child Health, 38th and R Streets, NW, Washington, D.C., 20057. (202) 625-8400.

A Directory of Voluntary Organizations in Maternal and Child Health. 1987. Washington, D.C.: National Center for Education in Maternal and Child Health.

This 39-page directory includes but is not limited to voluntary organizations that deal with genetic diseases. The organizations listed perform a variety of services—publishing educational materials, disseminating general information, making referrals, and furnishing support for individuals with specific needs. The directory is divided into five sections: (1) voluntary organizations in the United States, listed alphabetically by disease; (2) self-help clearinghouses listed by state; (3) voluntary organizations in Canada; (4) selected societies and professional associations; and (5) an index.

The directory is available free of charge from the National Center for Education in Maternal and Child Health, 38th and R Streets, NW, Washington, DC, 20057. (202) 625-8400.

SELECTED ORGANIZATIONS

National Center for Education in Maternal and Child Health. The center provides educational services to organizations, agencies, and individuals with an interest in maternal and child health. The Center's staff responds to requests for information, provides technical assistance, maintains a resource center, and develops publications on maternal and child health topics and resources. For more information, contact the National Center for Education in Maternal and Child Health, 38th and R Streets, NW, Washington, DC, 20057. (202) 625-8400.

March of Dimes Birth Defects Foundation. This voluntary health organization exists for the prevention of genetic and nongenetic birth defects and their consequences. It develops and distributes educational materials for health professionals and the public, works with other national and local organizations to initiate and implement community programs of prenatal care education and service, makes basic and clinical research grants and medical service grants, and sponsors medical conferences and symposia. The national headquarters of the March of Dimes Birth Defects Foundation is located at 1215 Mamaroneck Avenue, White Plains, New York, 10605. (914) 428-7100.

Council of Regional Networks of Genetic Services (CORN). This organization exists to promote communication among the regional genetic services networks and between the networks and other professional organizations, public agencies, and voluntary groups. For information, contact Dr. R. Stephen Amato, President, Council of Regional Networks for Genetic Services, Chairman, Department of Pediatrics, Greater Baltimore Medical Center, 6701 N. Charles Street, Baltimore, MD 21204. (301) 828-2780.

The American Society of Human Genetics. The main objectives of this professional society are to promote contact among investigators in human genetics and to encourage and integrate research in human genetics. The society publishes a monthly scientific journal, *The American Journal of Human Genetics.* This is a technical journal not oriented toward the lay public. For more information, contact the American Society of Human Genetics, 9650 Rockville Pike, Bethesda, MD 20814. (301) 571-1825.

Acknowledgments

The meeting and workshop on which this book is based were sponsored jointly by the Institute of Medicine and the National Academy of Sciences, which appointed a committee that was instrumental throughout the course of the planning of the meeting and preparation of the book. It is a pleasure to acknowledge the leadership provided by LEON E. ROSENBERG, Dean and C. H. N. Long Professor of Human Genetics, Medicine, and Pediatrics, Yale University School of Medicine, who chaired the planning committee and the annual meeting program, and by THEODORE FRIEDMANN, Professor, Department of Pediatrics, University of California at San Diego, who chaired the workshop. It is also a pleasure to acknowledge the significant contributions of BARBARA FILNER, Director of the Institute of Medicine Division of Health Sciences Policy, who was responsible for planning and implementing the program and book.

Members of the planning committee and speakers at the meeting and workshop, all of whom graciously cooperated throughout the preparation of this book, are listed below.

SPEAKERS AND PLANNING COMMITTEE MEMBERS

W. FRENCH ANDERSON, Chief, Laboratory of Molecular Hematology, National Heart, Lung, and Blood Institute, National Institutes of Health

MARIO R. CAPECCHI, Professor, Department of Biology, University of Utah

ROBERT J. DESNICK, Professor, Pediatrics and Genetics, and Chief, Division of Medical Genetics, Mount Sinai School of Medicine

JAMES D. EBERT, President, Carnegie Institution of Washington

HUNG FAN, Associate Professor, Department of Molecular Biology and Biochemistry, and Director, Cancer Research Institute, University of California at Irvine

ELI GILBOA, Associate Member, Memorial Sloan-Kettering Cancer Center

ROCHELLE HIRSCHHORN, Professor, Department of Medicine, New York University School of Medicine

RUDOLF JAENISCH, Member, Whitehead Institute for Biomedical Research, and Professor of Biology, MIT

WILLIAM N. KELLEY, John G. Searle Professor and Chairman, Department of Internal Medicine, University of Michigan Medical Center

PHILIP LEDER, John Emory Andrus Professor and Chairman, Department of Genetics, Harvard Medical School, and Senior Investigator, Howard Hughes Medical Institute

MAURICE J. MAHONEY, Professor of Human Genetics, Pediatrics, Obstetrics and Gynecology, Department of Human Genetics, Yale University School of Medicine

DAVID W. MARTIN, JR., Vice President for Research, Genentech, Inc.

A. DUSTY MILLER, Assistant Member, Department of Molecular Medicine, Fred Hutchinson Cancer Research Center, Seattle

ARNO G. MOTULSKY, Professor of Medicine and Genetics and Director, Center for Inherited Diseases, University of Washington

RICHARD MULLIGAN, Member, Whitehead Institute for Biomedical Research, and Associate Professor of Biology, MIT

DANIEL NATHANS, Professor of Molecular Biology and Genetics, Johns Hopkins University School of Medicine

STUART H. ORKIN, Leland Fikes Professor of Pediatric Medicine, Harvard Medical School

ROBERTSON PARKMAN, Head, Division of Research Immunology and Bone Marrow Transplantation, Childrens Hospital of Los Angeles

MAXINE F. SINGER, Chief, Laboratory of Biochemistry, National Cancer Institute, National Institutes of Health

WILLIAM M. SUGDEN, Professor of Oncology, McArdle Laboratory for Cancer Research, University of Wisconsin

HAROLD E. VARMUS, Professor, Department of Microbiology and Immunology, University of California at San Francisco

INDER M. VERMA, Professor, Salk Institute for Biological Studies

LeRoy Walters, Director, Center for Bioethics, Joseph and Rose Kennedy Institute of Ethics, Georgetown University
James B. Wyngaarden, Director, National Institutes of Health

Also deserving of special thanks for their contributions to this publication are the following: Mary Ampola, Director, Amino Acid Laboratory, The Floating Hospital for Infants and Children at New England Medical Center; Robin J. R. Blatt, Genetics Specialist, Massachusetts Genetics Program, Massachusetts Department of Health; Roscoe O. Brady, Branch Chief, Developmental and Metabolic Neurology Branch, National Institute of Neurological and Communicative Disorders and Stroke, National Institutes of Health; Iva H. Cohen, Manager, Human Gene Mapping Library, Howard Hughes Medical Institute, New Haven; Robert M. Cook-Deegan, Senior Analyst, Office of Technology Assessment, Congress of the United States; Frederick Doherty, Director, Ultrasound Division, New England Medical Center; Walter Dzik, Director, Blood Bank and Tissue Typing Laboratory, New England Deaconess Hospital; Michael S. Hershfield, Associate Professor, Division of Rheumatology and Immunology, and Assistant Professor, Department of Biochemistry, Duke University Medical Center; Jan Hudis, Science Information Editor, March of Dimes Birth Defects Foundation; Joan King, Associate Professor, Department of Anatomy and Cellular Biology, Tufts University School of Medicine; Katherine Wood Klinger, Manager of Genetic Disease Research, Integrated Genetics; Marianne Laveille, Research Associate, Institute of Medicine; M. Louise Markert, Assistant Professor, Department of Pediatrics, Duke University Medical Center; Robert Matousek, Biostatistician, March of Dimes Birth Defects Foundation; R. Scott McIvor, Assistant Professor, Institute of Human Genetics, Department of Laboratory Medicine and Pathology, University of Minnesota; Henry I. Miller, Special Assistant to the Commissioner, Food and Drug Administration; Patricia A. Ward, Genetic Counselor and Coordinator, Kleberg Prenatal Diagnostic Center, Baylor College of Medicine; James M. Wilson, Visiting Scientist, Whitehead Institute for Biomedical Research.

The challenging task of writing the book based on the 1986 Institute of Medicine annual meeting and workshop was accomplished with skill and grace by Eve K. Nichols.

The writing of this manuscript was made possible by the National Research Council Fund, a pool of private discretionary, nonfederal funds used to support programs initiated by the National Academy of Sciences that are concerned with national issues in which science and technology figure significantly.

Samuel O. Thier
President
Institute of Medicine

Glossary

ADA DEFICIENCY: *see* Adenosine deaminase deficiency.

ADENINE: One of the nucleotide bases—constituents of the chemical building blocks that make up DNA and RNA. Abbreviated by the letter A.

ADENOSINE DEAMINASE (ADA) DEFICIENCY: A genetic disease, inherited as an autosomal recessive trait, caused by deficiency of the enzyme adenosine deaminase. Patients are extremely susceptible to a broad range of infections. Unless special measures are taken, ADA-deficient children usually die before age 2. ADA deficiency causes about 20 percent of all cases of severe combined immune deficiency (SCID).

ADRENOLEUKODYSTROPHY: An X-linked recessive disorder marked by rapidly progressive deterioration of the white matter of the brain in childhood. Symptoms include dementia, blindness, and limb weakness, associated with adrenal insufficiency.

AFP: *see* Alpha-fetoprotein.

AGAMMAGLOBULINEMIA: Absence or severe depression of one or more types of antibody—proteins in the body that recognize and bind to foreign substances. The most common form of inherited agammaglobulinemia is an X-linked recessive disease in which affected males are extremely vulnerable to certain types of bacterial infections. The severe, recurrent infections can be prevented by regular injections of gamma globulin.

ALPHA-FETOPROTEIN: A protein present in appreciable amounts in the blood of the fetus, infant, and normal pregnant woman. Women carrying fetuses with spina bifida and related defects of the central nervous system have higher levels of AFP in their blood (and in their amniotic fluid) than women carrying healthy fetuses. Pregnancies in which the fetus has Down syndrome are associated with lower-than-expected AFP levels in the mother's blood.

ALPHA₁-ANTITRYPSIN DEFICIENCY: A genetic disease, inherited as an autosomal recessive trait, caused by deficiency of the protein alpha₁-antitrypsin. The deficiency causes severe lung disease in adults and, less commonly, life-threatening liver disease in infants and children.

ALPHA THALASSEMIA: *see* Thalassemia

ALPORT SYNDROME: A hereditary disease associated with both deafness and progressive kidney disease. The mode of inheritance may be either autosomal dominant or X-linked.

AMINO ACID: The chemical building blocks of protein. Twenty different amino acids are found in the proteins of human beings.

AMINO ACID DISORDER (AMINOACIDOPATHY): Any of a large number of inherited or acquired diseases involving the metabolism or transport of amino acids. Examples of inherited aminoacidopathies include citrullinemia, homocystinuria, ornithine transcarbamylase deficiency, phenylketonuria, and tyrosinemia.

AMNIOCENTESIS: A procedure in which a physician inserts a needle through the abdominal wall of a pregnant woman to withdraw a small amount of fluid from the amniotic sac (the fluid-filled sac surrounding the fetus). It is usually performed after the 14th week of pregnancy. Analysis of fetal cells present in the fluid can be used to detect chromosomal abnormalities and many inborn errors of metabolism. Spina bifida and other neural tube defects can be identified by an increased concentration of alpha-fetoprotein in the amniotic fluid.

ANENCEPHALY: A congenital defect in which part of the brain (the cerebral hemisphere) is missing or greatly reduced in size.

ANTIBODY: A protein in the blood produced in response to exposure to specific foreign molecules. Antibodies neutralize toxins and interact with other components of the immune system to eliminate infectious microorganisms from the body.

AUTOSOME: Any of the chromosomes except the sex chromosomes (the X and Y chromosomes). Human beings have 22 pairs of autosomes.

AUTOSOMAL DOMINANT INHERITANCE: For a genetic disease, a pattern of inheritance in which one copy of a mutant gene on an autosomal chromosome is sufficient to cause disease. Each child of an affected individual has a 50 percent chance of inheriting the mutant gene (and, therefore, of developing the disease). Sons and daughters are affected equally. Unaffected individuals do not transmit the trait.

AUTOSOMAL RECESSIVE INHERITANCE: For a genetic disease, a pattern of inheritance in which a mutant gene must be present on both members of a pair of autosomal chromosomes for disease to occur. If two people carry the same autosomal recessive trait, each of their offspring will have a 25 percent chance of inheriting two copies of the mutant gene and, therefore, of developing the disease. Sons and daughters are affected equally.

BACTERIOPHAGE: A virus that infects bacterial cells.

BASE: In DNA, one of four molecules (adenine, cytosine, guanine, or thymine) that contribute the informational content to nucleotide building blocks. In RNA, uracil substitutes for thymine.

BASE PAIR: Two bases, one on each strand of a double-stranded DNA molecule, that are attracted to each other by weak chemical interactions. In DNA, adenine pairs only with thymine and guanine pairs only with cytosine. The sequence of one strand of a double-stranded DNA molecule can be deduced by knowing the sequence of its partner. This complementarity is the key to the self-replication and information transmitting capabilities of DNA. (When the information in DNA is transcribed into RNA, uracil replaces thymine.)

BETA GLOBIN: *see* Hemoglobin.

BETA THALASSEMIA: *see* Thalassemia.

BIOTINIDASE DEFICIENCY: An autosomal recessive disease that can cause poor growth, developmental delay leading to mental retardation, skin rash, hair loss, and metabolic acidosis. Death in infancy also has been reported. The disease is caused by an inherited defect in biotinidase, an enzyme involved in the metabolism of biotin, an important vitamin. Early detection through newborn screening allows preventive treatment with biotin supplementation.

BLOOD–BRAIN BARRIER: The functional barrier that prevents the passage of enzymes and other molecules from the blood into the central nervous system.

B LYMPHOCYTE: A type of white blood cell that produces antibody.

BONE MARROW: The soft pulpy material in the cavities of bones. The marrow contains blood-forming cells that give rise to red blood cells, platelets, white blood cells and other elements of the immune system, and cells that contribute to bone growth.

BONE MARROW TRANSPLANT: Transfer of bone marrow cells from one person (the donor) to another (the recipient). To reduce the likelihood of a life-threatening immune reaction, bone marrow transplantation for genetic diseases usually is limited to patients who have genetically matched siblings. (*See* Histocompatibility.)

CARBOHYDRATE: A class of compounds that includes both small sugar molecules, such as glucose, fructose, and galactose, and larger molecules that consist of chains of the simple sugars (for example, starch and glycogen).

CARDIOMYOPATHY: Disease of the muscular walls of the heart.

CARRIER: A person who has one copy of a gene associated with a recessive genetic disease and one copy of its normal counterpart. The carrier usually does not have symptoms of the disease, because the protein product of the normal gene is sufficient for normal function.

CARRIER SCREENING: Tests to determine whether a person carries a gene associated with a genetic disease (the tests may involve enzyme measurements or direct examination of DNA). Carrier screening for a disease usually is carried out in specific populations known to be at increased risk—for example, screening for Tay-Sachs disease among Jews of European ancestry or for sickle cell disease among blacks. Other candidates for carrier screening include couples who have had a child with a genetic disease or the siblings of an affected child.

cDNA (complementary DNA): DNA that is synthesized in a test tube using messenger RNA as the template for the ordering of bases.

CELL: The basic subunit of any living organism; the simplest unit that can exist as an independent living system. The cell consists of a nucleus (containing the chromosomes) and various organelles for special functions (such as mitochondria for energy metabolism) in a matrix surrounded by a membrane.

CELL-MEDIATED IMMUNITY: The part of the immune system primarily responsible for protection against viral and fungal infections and for the rejection of transplanted organs.

CENTRAL NERVOUS SYSTEM (CNS): The brain and spinal cord.

CHEDIAK-HIGASHI SYNDROME: A genetic disease, inherited as an autosomal recessive trait, that is characterized by recurrent infections, enlargement of the liver and spleen, partial albinism, central nervous system abnormalities, and a high incidence of cancer of the lymphoid tissues. Most affected individuals die during childhood.

CHORIONIC VILLUS SAMPLING (CVS): A procedure performed in the ninth through eleventh weeks of pregnancy to obtain cells for prenatal diagnosis. Physicians collect a small sample of chorionic villi (fingerlike projections of the membrane surrounding the embryo early in pregnancy). Chorionic villus cells carry the same genetic information as the developing fetus. Analysis of the cells by conventional methods or by recombinant DNA techniques can be used to diagnose many genetic diseases.

CHROMOSOME: A structure in the cell nucleus that contains the hereditary material. Chromosomes are composed of DNA and proteins; they can be seen with the light microscope during certain stages of cell division. Genes are arranged in a linear fashion along the chromosomes.

CHROMOSOMAL DISORDER: A genetic disease involving an irregularity in the number or composition of chromosomes. The most familiar chromosomal disorder is Down syndrome. (*See also* Trisomy and Translocation.)

CHRONIC GRANULOMATOUS DISEASE: A genetic disorder usually transmitted as an X-linked inherited trait. Impaired function of phagocytic cells in the body leads to severe, repeated infections of the skin, lymph nodes, liver, lungs, and bones. The disorder was once uniformly fatal in childhood, but better supportive medical care has improved the outlook for these patients.

CITRULLINEMIA: An autosomal recessive disease caused by a deficiency of the urea-cycle enzyme argininosuccinate synthetase. (The urea cycle is the way the body disposes of waste nitrogen.) Prior to the development of current therapeutic techniques, patients exhibited mental retardation, vomiting, and convulsions; coma and death usually occurred in the newborn period. Long-term survival is now possible in some cases with a low protein diet and appropriate amino acid supplements.

CLONING: *See* Gene cloning.

CLONING VECHICLE: A small plasmid, phage, or animal virus DNA molecule used to transfer a fragment of DNA into a living cell. Cloning vehicles are capable of multiplying inside living cells. Thus, if a cloning vehicle transfers a specific fragment of DNA into a cell that is also multiplying, all the progeny of the cell will contain identical copies of the vehicle and the transferred DNA fragment.

CODON: The "word" in the language of the genetic code. A codon consists of three adjacent "letters" (nucleotides) in DNA or RNA. Each codon corresponds to one of 20 amino acids or to a signal to start or stop the construction of an amino acid chain.

CONGENITAL: Present at birth. Some congenital conditions are inherited, some are caused by environmental factors, and some are caused by a combination of hereditary and environmental factors.

CONGENITAL ADRENAL HYPERPLASIA (CAH): A family of genetic disorders associated with abnormal development of the genitalia in females and early puberty in males. The most common form of the disorder is caused by deficiency of the enzyme 21-hydroxylase. If the enzyme deficiency is severe, an affected infant may experience a life-threatening medical crisis (with signs of adrenal insufficiency) in the first two weeks of life. Therapy with corticosteroids prevents such a crisis, suppresses excessive adrenal androgen production, averts further virilization, and allows a normal onset of puberty. Most forms of CAH are believed to be transmitted as autosomal recessive traits.

CRETINISM: A syndrome involving irreversible mental retardation, deafness, and abnormal skeletal growth that results from a lack of thyroid hormone during the first year or two of life. The condition may arise from a defect in any one of the many enzymes involved in the production, storage, secretion, delivery, and utilization of the thyroid hormones. The mode of inheritance is usually autosomal recessive. Cretinism can be corrected and controlled by the prompt administration of thyroid hormone, as long as irreparable damage to the central nervous system has not already occurred.

CRIGLER-NAJJAR SYNDROME: A genetic disease in which the liver enzyme bilirubin UDP-glucuronosyl transferase is absent. Waste products from the normal breakdown of red blood cells accumulate in the blood, leading to severe damage to the central nervous system. Most children with this autosomal recessive disease die in infancy or early childhood. In Crigler-Najjar syndrome, type 2, there is a partial deficiency of bilirubin UDP-glucuronosyl transferase. Most of those affected survive to adulthood without neurological problems. The mode of inheritance is uncertain.

CYSTIC FIBROSIS: A genetic disease in which a high-protein viscous material interferes with the normal function of glands throughout the body. The lungs, pancreas, and sweat glands are most prominently affected. Effective treatment of lung complications now permits the survival of many patients into adulthood.

CYSTINOSIS: Three related genetic disorders associated with defective metabolism of the chemical cystine. Cystine crystals are deposited in many body tissues. In the early-onset form, children grow poorly and develop extreme sensitivity to light within the first few years of life. Kidney failure usually occurs before age 10. In the benign form of cystinosis, cystine crystals develop but do not cause disease. An intermediate form of cystinosis has variable symptoms. Kidney transplantation is used to treat cystinosis patients with terminal kidney disease. All three forms of cystinosis appear to be transmitted as autosomal recessive traits.

CYSTINURIA: An inherited disorder of amino acid transport that affects the kidneys and the gastrointestinal tract. Patients develop stones in the urinary tract. The disease often appears first between 10 and 30 years of age. The variants of this disorder all appear to be transmitted as autosomal recessive traits.

CYTOPLASM: All of the contents of a cell except the nucleus.

CYTOSINE: One of the nucleotide bases—constituents of the chemical building blocks that make up DNA and RNA. Abbreviated by the letter C.

DNA (DEOXYRIBONUCLEIC ACID): The molecule that contains the hereditary information in all organisms except the RNA viruses. Its chemical structure is that of a double helix—two long strands twisted around one another. Each strand is a linear chain of nucleotides; the sequence of nucleotides specifies the structure of all RNA molecules (their nucleotide sequences) and protein molecules (their amino acid sequences) made by the cell. The two strands of DNA are held together by weak chemical bonds (hydrogen bonds) between complementary base pairs (adenine always pairs with thymine and guanine always pairs with cytosine).

DOWN SYNDROME: A chromosomal disorder caused by an extra copy of chromosome number 21. Three of the most serious features of Down syndrome are mental retardation, congenital heart defects (in about 40 percent of cases), and increased susceptibility to infection.

DEFECTIVE VIRUS: A virus with an incomplete or incapacitated genome that is not capable of reproducing itself inside a cell.

DUCHENNE MUSCULAR DYSTROPHY: An X-linked genetic disorder associated with progressive muscle weakness beginning at about 3 years of age. Affected boys eventually become unable to walk. Death usually occurs as a result of cardiac failure or respiratory infection by age 20.

ELECTROPORATION: A procedure used to transfer foreign DNA into cells. An electrical current is applied to the cells, which then take up DNA and other chemicals from surrounding fluids.

ENHANCER: A short segment of DNA that influences the level of expression of genes adjacent to it by controlling the frequency with which transcription is initiated.

ENZYME: A protein that speeds up chemical reactions in a cell by acting as a catalyst. (Almost all chemical reactions in the cell—production of energy, replication of DNA, synthesis of protein, digestion of food molecules, and breakdown of waste products—require enzymes.)

ENZYME REPLACEMENT THERAPY: Supplying an enzyme that the body fails to produce because of a genetic mutation.

EXON: The segment of DNA in a gene that encodes the structure of a protein molecule. Many genes in animals and plants are composed of exons separated by intervening sequences of DNA called introns. The whole gene is copied into messenger RNA, but the introns are excised and the exons are spliced together before the messenger RNA is translated into protein.

EXPRESSION: The activation of information in a gene. The DNA is copied into messenger RNA (in a process called transcription), and the messenger RNA then directs the synthesis of a protein molecule (in a process called translation).

FABRY DISEASE: An X-linked recessive disease caused by deficiency of the enzyme alpha-galactosidase A. The enzyme's substrate (the substance on which it acts) accumulates in the lysosomes of blood vessels, leading to skin lesions, kidney failure, and cardiovascular disease. The most common childhood symptom is recurrent fever in association with severe pain of the hands and feet. Death most often results from kidney failure; the average age at death is about 40, although some patients have lived into their sixties.

FACTOR VIII: A protein necessary for blood clotting. Deficiency of Factor VIII causes hemophilia A.

FACTOR IX: A protein necessary for blood clotting. Deficiency of Factor IX causes hemophilia B.

FAMILIAL HYPERCHOLESTEROLEMIA: A genetic disease in which lipoprotein metabolism is defective (because of one of several mutations in the gene specifying the cell surface receptor for low density lipoprotein), leading to high blood cholesterol levels. The pattern of inheritance is autosomal dominant—individuals may be heterozygous or homozygous for the mutant gene. Heterozygotes have one normal gene and one mutant gene for the LDL receptor. They have moderately elevated cholesterol levels from birth and develop tendon xanthomas (orange or yellow lipid deposits in the tendons) and coronary artery disease after age 30. In homozygotes, xanthomas of the skin and tendons appear within the first 4 years of life and coronary artery disease begins in childhood. Angina, heart attacks, and sudden death occur commonly in homozygotes between the ages of 5 and 30. Familial hypercholesterolemia is one of the most common inborn errors of metabolism; the frequency of heterozygotes in the U.S. population is about 1 in 500. The frequency of homozygotes is much less—about 1 in one million.

FANCONI ANEMIA: A genetic disease involving a deficiency in bone development and bone marrow function. It is transmitted as an autosomal recessive trait.

FANCONI SYNDROME: A characteristic feature of several genetically transmitted metabolic diseases, including cystinosis, hereditary fructose intolerance (when fructose is not restricted), galactosemia (when galactose is not restricted), and Wilson disease. The syndrome consists of two components: disruption of kidney function and a metabolic bone disease. Fanconi syndrome is reversible in hereditary fructose intolerance, Wilson disease, and other diseases in which the underlying disorder can be treated, but it is not reversible in cystinosis.

FETAL THERAPY: Treatment of a fetus prior to birth, either directly—with intrauterine transfusions or surgery—or indirectly by the administration of certain drugs to the mother (such as vitamins or antiarrhythmic agents).

FETOSCOPY: A procedure that permits viewing of the fetus in the womb by means of a fiber optic instrument.

FIBROBLAST: The principal cell of connective tissue in the body. The main function of fibroblasts is to produce collagen and other constitutents of the gelatinous mixture that binds cells together.

FRAME-SHIFT MUTATION: A change in the information encoded in a gene resulting from the insertion or deletion of one or two nucleotides. Such a mutation establishes a new reading frame (regrouping of nucleotide triplets)—every codon downstream from the mutation site is altered.

FRUCTOSE INTOLERANCE: An autosomal recessive disease in which the liver enzyme fructose-1-phosphate aldolase is deficient. Consumption of fructose (in fruits or table sugar, for example) causes vomiting, failure to thrive, disruption of liver function, and other symptoms. The symptoms can be avoided by eliminating fructose from the diet.

GALACTOSEMIA: An autosomal recessive disease in which deficiency of the enzyme galactose-1-phosphate uridyl transferase results in an inability to digest galactose (a sugar found in dairy products). Symptoms include vomiting, liver disease, cataracts, and mental retardation. Elimination of galactose from the diet causes a striking regression of symptoms, and children who are treated early may develop normally. However, the diet cannot reverse damage that occurs before birth. Some children with galactosemia have learning disabilities, speech defects, and, in girls, ovarian abnormalities, despite early treatment.

GAMMA GLOBULIN: A class of blood proteins that includes antibodies, molecules that participate in the immune response.

GAUCHER DISEASE: A genetic disease resulting from deficiency of the enzyme lysosomal glucocerebrosidase. The enzyme's substrate (the substance on which it acts) accumulates in large white blood cells in the bone marrow, spleen, and liver, causing bleeding disorders and enlargement of the affected organs. Gaucher disease occurs in three forms: (1) a chronic adult type that does not affect the central nervous system, (2) an acute infantile type that is usually apparent before six months of age, destroys the central nervous system, and causes death by age two, and (3) a juvenile type in which the central nervous system is involved, but the signs of neurologic damage are less severe and appear later than in type 2 patients. All three types are inherited as autosomal recessive traits.

GENE: A sequence of DNA that contains information for the construction of one protein molecule. (In classical genetics, the fundamental unit of heredity that carries a single trait.)

GENE SPLICING: The insertion of a piece of DNA into another DNA molecule in the test tube.

GENE CLONING: A way of using microorganisms to produce millions of identical copies of a specific piece of DNA.

GENE TRANSFER: Insertion of a foreign gene into a cell in such a way that it will remain stable (will not be destroyed by the cell) and will be passed on to descendants of the cell during cell division.

GENETIC DISEASE: A disease that is due to an abnormality in the information content of a person's DNA.

GENETIC ENGINEERING: The manipulation of the DNA of a cell or organism to change specific characteristics.

GENOME: The total genetic information contained in an organism's genes.

GERM LINE GENE THERAPY: The insertion of a normal gene into the fertilized egg of an animal that has a specific genetic defect. Every cell in the body acquires the new gene, including the reproductive cells, so it is passed on to succeeding generations.

GLUCOSE-6-PHOSPHATE DEHYDROGENASE (G-6-PD) DEFICIENCY: The most common disease-producing enzyme deficiency in human beings—about 100 million people in the world are affected. The disorder is inherited as an X-linked trait. In the United States, about 1 in 10 black men have a relatively mild form of G-6-PD deficiency referred to as the A type. G-6-PD is important for the normal maintenance of red blood cells. People with G-6-PD deficiency develop hemolytic anemia (anemia resulting from the premature destruction of red blood cells) after treatment with certain drugs (especially antimalarial agents and sulfonamides) or after the ingestion of a chemical found in fava beans. Infections and diabetic acidosis also can cause hemolysis. In mild cases, the only symptom may be darkening of the urine; more severe cases are associated with weakness, abdominal and back pain, jaundice, and black urine. The management of patients with G-6-PD deficiency consists primarily of avoidance of drugs known to be hemolytic.

GLYCOGENOSES (GLYCOGEN STORAGE DISEASES): A group of genetic diseases caused by a deficiency of one of the enzymes involved in the degradation of glycogen, the principal form in which carbohydrate is stored in the body. Examples include von Gierke disease, Pompe disease, and Hers disease.

GLYCOPROTEIN: A sugar-protein complex.

GRAFT VERSUS HOST DISEASE (GVHD): A possible complication of bone marrow transplantation. White blood cells in the donor bone marrow recognize cells of the transplant recipient as foreign and attack them, injuring the skin, the liver, the intestinal tract, and other tissues.

GUANINE: One of the nucleotide bases—constituents of the chemical building blocks that make up DNA and RNA. Abbreviated by the letter G.

HELPER CELL: A cell that has been altered genetically to allow the production of a replication-defective virus capable of infecting specific target cells. Also called a packaging cell.

HELPER VIRUS: A defective retrovirus inserted into the genome of a helper cell to provide functions missing from the DNA of a retroviral vector. The vector DNA cannot make the enzymes and other proteins necessary to form a complete viral particle, because the necessary viral genes have been replaced by foreign genes. The helper virus DNA also is incomplete, but its defects are different from those of the vector DNA.

HEMATOPOIETIC: Relating to the formation of blood cells.

HEMOGLOBIN: The molecule in red blood cells that carries oxygen. Chemically, hemoglobin consists of four protein chains and an iron-containing structure called a heme group. The major component of normal adult hemoglobin is hemoglobin A, which contains two alpha-globin chains and two beta-globin chains.

HEMOPHILIA: A group of X-linked inherited diseases that involve defects in the blood-clotting proteins. Hemophilia A, the most common form of hemophilia, is caused by a deficiency of factor VIII. Hemophilia B is caused by a deficiency of factor IX.

HERS DISEASE: An autosomal recessive disorder due to deficiency of the enzyme liver phosphorylase. It is a glycogen storage disease. Symptoms are relatively mild and include enlargement of the liver, increased susceptibility to infection, and bleeding tendencies.

HETEROZYGOUS: Having two different forms of a given gene on the paired chromosomes of a cell. When this occurs, an individual is said to be heterozygous for that gene. If the two copies of a given gene are identical, the individual is said to be homozygous for that gene.

HISTIDINEMIA: An autosomal recessive disease caused by lack of the enzyme histidase. Affected individuals have excessive amounts of histidine in their blood and urine. The majority of children are normal; some have been reported with speech defects and mental retardation.

HISTOCOMPATIBILITY: Sufficient similarity between two individuals so that tissue from one will not be recognized as foreign and rejected by the other. The acceptance or rejection of transplanted tissue depends on the presence or absence of specific histocompatibility antigens on the cells of the transplanted tissue. These proteins enable immune cells to distinguish between "self" and "nonself." Bone marrow transplantation for genetic diseases usually is limited to patients who have a sibling that has inherited the same major histocompatibility antigens.

HOMOCYSTINURIA: A group of genetic disorders associated with excessive excretion of homocystine in the urine. Symptoms include mental retardation, dislocation of the lenses of the eyes, skeletal abnormalities, and excessive blood clotting. Early treatment (involving a restricted diet and, in some cases, vitamin supplementation) can prevent or reduce the symptoms.

HOMOZYGOUS: Having two identical copies of a given gene on the paired chromosomes of a cell. When this occurs, an individual is said to be homozygous for that gene. (*See also* Heterozygous.)

HPRT: The enzyme hypoxanthine-guanine phosphoribosyltransferase. The total absence of HPRT causes the genetic disease Lesch-Nyhan syndrome. Partial deficiency of HPRT leads to gout. HPRT deficiencies are inherited as X-linked recessive traits.

HUMORAL IMMUNITY: The human defense mechanism that involves the production of antibodies.

HUNTINGTON DISEASE: A genetic disease characterized by uncontrollable physical movements, loss of speech, and progressive mental deterioration. The average age of onset is 38 years; death usually occurs within 10 to 15 years. The mode of inheritance is autosomal dominant.

HUNTER SYNDROME: A genetic disease caused by deficiency of the enzyme iduronate sulfatase. The most common features are joint stiffness, dwarfing, and coarse facial features. Progressive mental deterioration also occurs, but the severity varies. Mildly affected patients may lead relatively normal lives and survive into their sixties, while those with severe disease may die in their early teens. The disease is inherited as an X-linked recessive trait.

HURLER SYNDROME: An autosomal recessive disease caused by deficiency of the enzyme alpha-L-iduronidase. It is a severe, progressive disease; death usually occurs before age 10. Cells of the nervous system, endocrine glands, liver, bones, and heart are most frequently affected. Dwarfing and clouding of the cornea become evident by 2 or 3 years of age. The usual causes of death are respiratory infection and heart failure.

HYDROCEPHALUS: Enlargement of the ventricles, or spaces, in the brain resulting from an abnormal accumulation of cerebrospinal fluid. In the fetus and infant, hydrocephalus disrupts normal development of the brain and may cause mental retardation or early death. The condition can be treated in some cases.

HYPERGLYCINEMIA: A group of hereditary disorders in which glycine accumulates in body fluids. The disorders can be grouped into two categories: nonketotic hyperglycinemia and ketotic hyperglycinemia. Nonketotic hyperglycinemia is not treatable—patients usually are severely mentally retarded and die in early childhood. In contrast, patients with ketotic hyperglycinemia who are diagnosed early often can be treated successfully with dietary therapy.

HYPOTHYROIDISM: An abnormal decrease in the activity of the thyroid gland resulting in deficient secretion of thyroid hormones. Hypothyroidism in infancy causes the classic signs of cretinism, including delayed physical and mental development.

IDIOPATHIC HEMOCHROMATOSIS: A genetic defect of iron metabolism causing iron to build up in body tissues. Symptoms include cirrhosis of the liver, heart damage, diabetes, decreased function of the testes or ovaries, pigmentation, and arthritis. Death occurs unless iron is eliminated by periodic bloodletting. The mode of inheritance appears to be autosomal recessive.

INBORN ERROR OF METABOLISM: A disease caused by a specific genetic defect in the function or structure of a protein molecule, usually an enzyme.

INTRON: A segment of DNA in a gene that does not contain information dictating the structure of the gene's protein product. Introns, or intervening sequences, are copied into messenger

RNA along with coding sequences, but they are removed before the messenger RNA is translated into protein. (The coding sequences are called exons.)

KARYOTYPE: The systematic display of chromosomes from a single somatic cell after the chromosomes have been dyed and then photographed through a microscope.

KERATINOCYTE: The primary cells in the protective, outermost layer of the skin.

KLINEFELTER SYNDROME: A chromosomal disorder in males caused by one or more extra X chromosomes (the most common sexual karyotype for this syndrome is XXY). Affected men are infertile and have small testes and poorly developed secondary sex characteristics. Many are tall and have subnormal intelligence.

KRABBE DISEASE: An autosomal recessive disease caused by deficiency of the enzyme galactocerebroside beta-galactosidase. Symptoms include progressive mental retardation, paralysis, blindness, and deafness. They usually appear between 3 and 6 months of age. Most patients die before the age of two.

LACTASE DEFICIENCY: A condition in which partial or total deficiency of the enzyme lactase results in an inability to digest the milk sugar lactose. The most common form of lactase deficiency develops in late childhood or adolescence: individuals who had been able to tolerate milk gradually notice symptoms of diarrhea, bloating, and cramps after consuming milk or some milk products. Lactase deficiency also may occur as a rare autosomal recessive disorder in infants and young children.

LACTOSE INTOLERANCE: The condition caused by lactase deficiency.

LESCH-NYHAN SYNDROME: An X-linked genetic disease caused by complete absence of the enzyme HPRT. Symptoms include self-mutilation, mental retardation, spastic cerebral palsy, and urinary tract stones.

LINKAGE ANALYSIS: The search for traits that tend to be inherited together (presumably because their respective genes are close to one another on a chromosome).

LIPID: Any of a group of substances that are greasy to the touch and insoluble in water, but soluble in alcohol and other fat solvents. Lipids are easily stored in the body, serve as a source of fuel, and are an important constituent of cell structures.

LIPOSOME: A structure with a lipid membrane that can be filled with a specific substance and then used to transport that substance to the interior of a target cell (the liposome membrane fuses with the cell's membrane).

LONG TERMINAL REPEAT: The segments on the ends of a retroviral genome after it has been transcribed from RNA into DNA. Adjacent to or within the long terminal repeats are the regulatory signals required for reverse transcription, for insertion of viral DNA into the host cell genome, and for expression of the virus (transcription of the DNA into messenger RNA, and translation of the messenger RNA into proteins).

LYMPHOCYTE: One of two types of white blood cells that play crucial roles in the immune system (see B lymphocyte and T lymphocyte).

LYSOSOME: A small spherical structure in the cytoplasm of a cell that contains enzymes capable of breaking down large molecules, such as complex lipids and carbohydrates.

LYSOSOMAL STORAGE DISEASE: The term for a group of diseases that result from inherited deficiencies of certain lysosomal enzymes. Symptoms are caused by the build up inside cells of the particular substance that a defective enzyme is supposed to degrade. The symptoms depend on the nature of the stored material and the function of the tissues involved. Lysosomal storage diseases are divided into four categories: glycogenoses (glycogen storage diseases), mucolipidoses, mucopolysaccharidoses, and sphingolipidoses.

MACROPHAGE: A cell derived from a certain type of white blood cell (a monocyte) that performs various functions in the immune system, including engulfing and destroying bacteria and other disease-producing microorganisms.

MAPLE SYRUP URINE DISEASE (MSUD): An autosomal recessive disease resulting in the build-up of three essential amino acids. Vomiting, failure to thrive, and severe neurological symptoms begin soon after birth. An odor, described as that of maple syrup or curry, is often noted in the urine. Untreated infants become progressively more lethargic and most die; survivors have a high likelihood of mental retardation. Dietary treatment, if begun early, may eliminate symptoms and allow normal development.

MAROTEAUX-LAMY SYNDROME: An autosomal recessive disease caused by deficiency of the enzyme lysosomal arylsulfatase B. The disorder causes severe skeletal deformities, including dwarfism, clouding of the corneas, and heart problems. Intelligence is normal, but death often occurs before age 30.

MENDELIAN DISEASE: *See* Monogenic disease.

MESSENGER RNA: The RNA molecule that carries a copy of the information encoded in a gene from the nucleus (the site of DNA) to the cytoplasm, where it is translated into the structure (amino acid sequence) of a protein molecule.

METABOLIC MANIPULATION: Treatment of a genetic disease by (1) limiting the intake of a potentially toxic substance (for example, a molecule that cannot be broken down because of a specific enzyme defect), (2) depleting stores of a potentially toxic substance (using drugs or other measures), or (3) replacing a substance the body is unable to manufacture.

METACHROMATIC LEUKODYSTROPHY (MLD): An autosomal recessive disease caused by deficiency of the enzyme arylsulfatase A. It causes progressive degeneration of the central nervous system, resulting in impairment of motor function, dementia, and eventually death. The disease usually begins in the second year of life and may be fatal by age 10. However, onset also may occur in childhood or in adult life.

METHYLMALONIC ACIDEMIA: A condition associated with elevated levels of methylmalonic acid in the blood and urine. It is caused by one of several genetic diseases, all of which appear to have autosomal recessive patterns of inheritance. Symptoms include lethargy, repeated bouts of vomiting, severe metabolic abnormalities, and developmental retardation. Many affected children die within the first week of life. A protein-restricted diet may prevent life-threatening problems and allow normal development. In addition, some patients respond to large doses of vitamin B_{12}.

MICROINJECTION: Insertion of DNA into a cell via injection with a very fine glass needle.

MISSENSE MUTATION: A change in one nucleotide in a sequence of DNA that results in the wrong amino acid being built into a protein.

MONOGENIC DISORDER: A disease caused by a defect in a single gene. Scientists have identified almost 4,000 monogenic disorders. They generally show one of three patterns of inheritance: autosomal recessive, autosomal dominant, or X-linked.

MSUD: *See* Maple syrup urine disease.

MUCOLIPIDOSES: A group of lysosomal storage diseases that lead to abnormal storage of both carboyhdrate molecules and lipid molecules.

MUCOPOLYSACCHARIDOSES: A group of lysosomal storage diseases involving enzymes responsible for degrading mucopolysaccharides (a class of carbohydrate molecules). Examples include Hurler syndrome, Hunter syndrome, Sanfilippo syndrome, and Maroteaux-Lamy disease.

MULTIPLE ENDOCRINE NEOPLASIA (MEN) SYNDROME: A group of several genetic disorders transmitted as autosomal dominant traits. The range of symptoms is extremely variable. It includes stomach ulcers, impaired kidney function, malignant tumors of the pancreas, thyroid cancer, vision disorders, and other problems.

MUTATION: A change in the structure of DNA that alters the information it encodes.

NONSENSE MUTATION: A mutation that leads to the production of a shortened protein molecule. The mutation changes a codon that had specified a particular amino acid in a protein chain into a stop signal.

NUCLEOTIDE: One of the building blocks of nucleic acids (DNA and RNA). Chemically, a nucleotide consists of three parts: a base, a sugar molecule, and a phosphate group.

NUCLEUS: The structure in a cell that contains the chromosomes.

ONCOGENE: A gene that is associated with the process of changing a normal cell into a cancerous cell.

ORGANIC ACID DISORDER (ORGANIC ACIDEMIA): Any of a large number of inherited or acquired diseases involving the metabolism of organic acid intermediates. Examples of inherited organic acid disorders include methylmalonic acidemia and propionic acidemia.

ORNITHINE TRANSCARBAMYLASE DEFICIENCY: An X-linked genetic defect associated with deficiency of the enzyme ornithine transcarbamylase. Symptoms include greatly increased levels of ammonia in the blood, recurrent vomiting, irritability, and central nervous system deterioration progressing to seizures, coma, or death. Long-term survival may be possible with a low protein diet and appropriate amino acid supplements.

OSTEOPETROSIS: A genetic disease marked by the formation of abnormally dense bone, presumably due to defects in normal bone remodeling. Symptoms include frequent fractures and bone deformities, anemia, neurological abnormalities, and facial paralysis. Bone marrow transplantation from a histocompatible donor has been used successfully to treat this disorder.

OXALOSIS: A general term for two diseases characterized by the excretion of large amounts of oxalic acid in the urine and by the formation of kidney stones and calcium deposits in the kidney. Symptoms of kidney disease usually begin before the age of 5, but there is variability in the age of onset and in the severity of symptoms. Death frequently occurs before the age of 20. The mode of inheritance of both disorders appears to be autosomal recessive.

PHAGE: *See* Bacteriophage.

PHAGOCYTE: Any cell that binds to, engulfs, and destroys bacteria or other foreign particles in the body.

PHENYLKETONURIA (PKU): An autosomal recessive disease caused by deficiency of the enzyme phenylalanine hydroxylase. The amino acid phenylalanine and some of its biochemical derivatives accumulate in the body and interfere with the normal development of brain cells, leading to mental retardation. Early identification of affected infants followed by dietary therapy permits normal or near-normal development.

PHOTOMICROGRAPH: A picture taken through a light microscope.

PLASMID: A small circular molecule of DNA. Plasmids occur naturally in many different types of bacteria. They remain separate from the single bacterial chromosome and they contain genes that are not essential for bacterial growth. Plasmids have the capability to reproduce themselves inside host cells.

POINT MUTATION: A change in DNA in which one nucleotide is replaced by another.

POLYCYSTIC KIDNEY DISEASE: A term used to describe two different disorders that both result in the formation of cysts in the kidneys. The cysts increase kidney size but reduce the amount of functional kidney tissue. Infantile polycystic kidney disease appears in infancy or childhood and is transmitted as an autosomal recessive trait; the earlier the onset, the more severe the disease. The adult form of the disease is transmitted as an autosomal dominant trait. High blood pressure and kidney failure develop as the condition progresses. Kidney transplantation has been successful in many cases.

POMPE DISEASE: An autosomal recessive disease caused by deficiency of the enzyme alpha-1,4-glucosidase. Glycogen accumulates in all tissues, leading to progressive muscle weakness.

In the severe infantile form, symptoms include congestive heart failure, and death usually occurs by one year of age. In two rarer forms of Pompe disease, symptoms appear later and progress more slowly, and heart disease is absent or minimal.

PORPHYRIA: A heterogeneous group of disorders associated with defects in the production of heme—the deep-red pigment that is a constituent of hemoglobin. The principal symptoms of the porphyrias are neurological abnormalities and skin photosensitivity. The neurological symptoms usually are linked to the ingestion of certain drugs. Most forms of porphyria are transmitted as autosomal dominant traits (the exception is congenital erythropoietic porphyria, which is a rare autosomal recessive disorder).

PRENATAL DIAGNOSIS: The diagnosis of genetic diseases or other disorders in a developing fetus.

PROBE: A short piece of DNA or RNA of known structure or function that has been tagged with some tracer substance (a radioactive isotope or specific dye-absorbing compound). It is used to locate and identify a specific gene or region of a chromosome.

PRODUCT REPLACEMENT: Treating a genetic disease by replacing the endproduct of a chemical reaction that cannot proceed normally because of a defective enzyme.

PROMOTER: A short sequence of DNA that controls the expression of a gene. It is the initial binding site for the cellular enzyme that copies DNA into RNA.

PROPIONIC ACIDEMIAS: A condition associated with elevated levels of propionic acid in the blood and urine. It is caused by one of several genetic diseases, all of which appear to have autosomal recessive patterns of inheritance. Symptoms include repeated bouts of vomiting, severe metabolic abnormalities, and developmental retardation. A protein-restricted diet may prevent life-threatening metabolic crises and allow normal development. In addition, some patients respond dramatically to biotin supplements.

PROTEIN: A large molecule that consists of one or more chains of amino acid building blocks. Proteins are essential to the structure and function of a cell.

PROTO-ONCOGENE: A gene capable of turning into an oncogene, a powerful inducer of unregulated (cancerous) growth. Under most conditions, proto-oncogenes participate in normal cell function. The insertion of viral control elements next to a proto-oncogene is one of the factors than can contribute to the abnormal activity of such a gene.

PROVIRUS: A copy of the genetic information of a retrovirus that is integrated into the DNA of an infected cell.

PURINE NUCLEOSIDE PHOSPHORYLASE (PNP) DEFICIENCY: An autosomal recessive disease caused by deficiency of the enzyme purine nucleoside phosphorylase. PNP-deficient children have a severe defect in cell-mediated immunity (but relatively normal humoral immunity), which makes them highly vulnerable to certain viral and fungal infections.

RADIOLABELING: The inclusion of a radioactive atom in the molecular structure of a chemical so that the chemical can be detected and measured by the presence of the radioactivity. (It is much easier to measure small amounts of radioactivity than small amounts of certain chemicals.)

RECOMBINANT DNA: A DNA molecule assembled in a test tube using segments of DNA from two different sources. The segments are cut with restriction enzymes and spliced together with an enzyme called DNA ligase.

REFSUM DISEASE: An autosomal recessive disease that results in the marked accumulation of phytanic acid in the blood and tissues. Symptoms include progressive loss of vision (especially night vision), degeneration of the peripheral nervous system, and disturbances in the coordination of voluntary movements. The time of onset ranges from early childhood to the mid-forties, but most patients have some signs of the disorder before age 20. The elimination of phytanic acid from the diet halts progression of the disease and may lead to restoration of some lost functions.

RESTRICTION ENZYME: An enzyme produced by a bacterial cell that recognizes and cuts a particular nucleotide sequence in DNA (the "recognition site" may range from 4 to 12 base pairs in length). The function of the enzyme in the bacterial cell is to protect the cell against the invasion of foreign DNA (for example, the DNA of a bacterial virus). Purified restriction enzymes are used by molecular geneticists to manipulate DNA in the test tube; they have played a crucial role in the development of recombinant DNA technology. Almost 600 different restriction enzymes are known.

RESTRICTION FRAGMENT LENGTH POLYMORPHISM: The variation in the length of pieces obtained when two similar DNA molecules are cut by a restriction enzyme; the variation results from a mutation that has increased or decreased the number of recognition sites in one of the pieces of DNA.

RETROVIRUS: One of a class of viruses that contain the genetic material RNA and that have the capability to copy this RNA into DNA inside an infected cell. The resulting DNA is incorporated into the genetic structure of the cell in the form of a provirus.

RETROVIRAL VECTOR: *See* Vector virus.

REVERSE TRANSCRIPTASE: The enzyme produced by retroviruses that allows them to make a DNA copy of their RNA. This is the first step in the virus's natural cycle of reproduction.

RNA (RIBONUCLEIC ACID): A class of molecules involved in converting the information contained in DNA into proteins. The nucleotide bases of the long, chainlike RNA molecules (usually found as a single strand) are adenine, cytosine, guanine, and uracil. Some viruses use RNA as their genetic material.

SANFILIPPO SYNDROME: A group of genetic diseases associated with defective degradation of the mucopolysaccharide heparan sulfate. The primary symptom is severe, progressive mental retardation, usually beginning in the first few years of life. Other symptoms include coarse facial features, enlargement of the liver, and joint stiffness, all of which are less pronounced than in other mucopolysaccharidoses. All forms of Sanfilippo syndrome are inherited as autosomal recessive traits.

SELECTABLE MARKER GENE: A gene that allows researchers to easily identify cells that have incorporated a vector virus. One example is the bacterial gene for neomycin resistance (*neo*). Expression of the *neo* gene by mammalian cells confers resistance to the drug G418, an analog of neomycin.

SERUM: The clear yellowish liquid that separates from blood when it is allowed to clot. It contains antibodies and other protein molecules.

SEVERE COMBINED IMMUNE DEFICIENCY (SCID): A group of genetic diseases in which both cellular immunity and humoral immunity (involving antibodies) are severely impaired. Patients with SCID are extremely susceptible to a broad range of infections. Before treatment was available, most patients died before the age of two. The current treatment of choice is a bone marrow transplant from a histocompatible donor.

SICKLE CELL DISEASE (ANEMIA): A genetic disease caused by the presence of a defective beta-globin chain in the hemoglobin molecule. Patients with sickle cell disease have red blood cells that tend to deform into a sickle-like shape when the abnormal hemoglobin crystalizes. The specific defect is caused by the replacement of one amino acid for another in the beta-globin chain (valine replaces glutamic acid). The major clinical symptoms include chronic anemia, impairment of growth and development, increased susceptibility to infection, leg ulcers, and painful "crises" caused by the obstruction of small blood vessels. In the U.S., most cases of sickle cell disease occur among blacks and Hispanics of Caribbean ancestry. The disease also affects some people of Arabian, Greek, Maltese, Sicilian, Sardinian, Turkish, and southern Asian ancestry.

SINGLE-GENE DISORDER: A genetic disease caused by a mutation in a single gene.

SOMATIC: Referring to all body tissues except the reproductive (germinal) tissues

SOMATIC CELL GENE THERAPY: A proposed technique for treating severe genetic diseases. A child with a life-threatening genetic disease, caused by a defect in a single gene, would be treated with the gene's normal counterpart. The normal gene, provided through recombinant DNA technology, would be inserted into a specific tissue in the child's body and would not be passed on to future generations.

SPHINGOLIPIDOSES: A group of lysosomal storage disorders caused by accumulation within the nervous system and other tissues of lipid molecules called sphingolipids. Examples include Fabry disease, Tay-Sachs disease, Krabbe disease, Gaucher disease, metachromatic leukodystrophy, and Niemann-Pick disease.

SPINA BIFIDA: A congenital defect in which the bony covering of the spinal cord fails to close during fetal development. It is believed to be caused by a combination of environmental and genetic factors.

SPLEEN: An organ in the abdominal cavity that removes worn-out red blood cells from the bloodstream, plays an important role in the immune system, and also serves as a reservoir for several types of blood cells.

STEM CELL: A cell that gives rise to the entire spectrum of blood-forming cells. Stem cells can replicate themselves to maintain the supply of stem cells, or they can differentiate along either of the two major pathways that leads to the formation of mature red and white blood cells.

SUBSTRATE: A molecule acted on by an enzyme.

SYNONYMOUS MUTATION: A type of point mutation (a change in a single nucleotide) that does not affect the protein product of a gene, because the altered codon codes for the same amino acid as did the original codon.

TARGET CELLS: In somatic cell gene therapy, cells removed from a patient with a severe genetic disease and then infected with a retroviral vector carrying a normal gene. When such cells are returned to the patient, they should carry the normal gene with them.

TAY-SACHS DISEASE: A genetic disease caused by deficiency of the enzyme hexosaminidase A. Affected infants exhibit developmental retardation, paralysis, dementia, and blindness beginning within the first four or five months of life. Death usually occurs before the age of 4 years. The gene defect occurs with greatest frequency among Jews of European ancestry. Mass screening programs have helped reduce the incidence of the disease.

THALASSEMIA: Any of several genetic diseases in which the decreased production of one hemoglobin subunit leads to an abnormal build-up of the other. The accumulation of excess globin chains disrupts the development of red blood cells and shortens their survival time, resulting in anemia. The severity of the anemia depends on the nature of the defect in the globin genes. The two most important disorders clinically are alpha thalassemia (deficient alpha-chain synthesis) and beta thalassemia (absent or deficient beta-chain synthesis).

THYMINE: One of the nucleotide bases—constituents of the chemical building blocks that make up DNA. Abbreviated by the letter T.

T-LYMPHOCYTE: A white blood cell that matures in the thymus gland and plays a variety of important roles in the immune system. T lymphtocytes are responsible for cell-mediated immunity, and they interact with B lymphocytes in the production of humoral immunity.

TRANSCRIPTION: A key step in the expression of a gene—the sequence of nucleotides in the gene is copied (transcribed) from DNA into a single-stranded molecule of messenger RNA.

TRANSDUCTION: The transfer of foreign DNA into a cell via a virus.

TRANSFECTION: The transfer of foreign DNA into a cell via chemical or physical means.

TRANSLATION: The process of decoding the information in a molecule of messenger RNA and using it to direct the construction of protein molecules specified in the messenger RNA.

TRANSLOCATION: A chromosomal aberration that occurs when a piece of one chromosome breaks off and attaches to another chromosome.

TRIPLE-X SYNDROME: A condition resulting from the presence of three X chromosomes. Affected women are generally fertile and have relatively normal sex differentiation. (Individuals with more than three X chromosomes are usually severely retarded.)

TRISOMY: The presence of one additional autosomal chromosome (three instead of two). The most common genetic disorders due to trisomy are trisomy 21 (Down syndrome), trisomy 18, and trisomy 13. The disorders all involve mental retardation, shortened life span, and structural abnormalities.

TURNER SYNDROME: A genetic disorder of females caused by abnormality of the sex chromosomes. Affected individuals usually have one X chromosome and no Y chromosome. Typical features include short stature, webbing of the neck, and failure of secondary sexual development. Intelligence is normal.

TYROSINEMIA (HEREDITARY): A group of autosomal recessive disorders in which large amounts of the amino acid tyrosine accumulate in the blood. Tyrosinemia, type 1, causes failure to thrive and severe, progressive liver disease. In tyrosinemia, type 2, the liver is normal but the deposition of tyrosine in body tissues causes skin and eye problems. Some patients are mentally retarded. Both forms of tyrosinemia respond to dietary therapy.

ULTRASOUND: The use of high-frequency sound waves to produce body images. Ultrasound is employed to locate the fetus and the placenta before procedures such as amniocentesis and chorionic villus sampling, and to diagnose a variety of structural abnormalities in the fetus.

URACIL: One of the nucleotide bases—a constituent of the chemical building blocks that make up RNA. Abbreviated by the letter U.

VECTOR VIRUS: A virus that has been altered with recombinant DNA technology to carry foreign DNA (for example, a human gene) into a cell. A vector constructed from a retrovirus provides the regulatory elements necessary for integration of the foreign DNA into the genome of the host cell and for expression of the DNA.

VIRION: A complete virus particle outside a cell.

VON GIERKE DISEASE: An autosomal recessive disease caused by deficiency of the liver enzyme glucose-6-phosphatase. The liver and kidneys of affected infants are enlarged with glycogen and fat deposits. Convulsions accompanying severe hypoglycemia (low blood sugar) may occur during the first year of life. Growth is poor and many patients have gout as young adults. Treatment includes frequent feedings and, in some cases, surgery.

WILSON DISEASE: A genetic disease in which excessive accumulation of copper in the liver and brain causes liver damage and a variety of neurological problems (neurologic effects are more frequent in adults). Treatment with the drug penicillamine, which removes copper from the blood, is very effective.

WISKOTT-ALDRICH SYNDROME: An X-linked disorder in which affected males have abnormal cellular immunity, abnormal platelet function and survival time, and decreased antibody response to certain antigens. Until recently, patients usually died of overwhelming viral or bacterial infections, bleeding, or tumors during the first decade. Bone marrow transplants from a histocompatible donor have been effective for many patients with this disease.

X-LINKED INHERITANCE: For a genetic disease, a pattern of inheritance in which a trait is passed from mother to son via an X chromosome. Each son of a woman carrying an X-linked disorder has a 50 percent chance of inheriting the abnormal gene—of developing the disease. Most X-linked traits are recessive, so daughters are not affected by the disease (they are protected by the gene's normal counterpart on the X chromosome inherited from their father). However, each daughter has a 50 percent chance of becoming a carrier like her mother.

Index

ADA. *See* Adenosine deaminase deficiency
Adenine, 24, 25, 26
Adenosine deaminase deficiency, 1, 2, 32, 38, 38n.2, 191; diagnosis, 46; treatment, 79–80, gene location, 100, 101; gene insertion in cells, 105, 135, 139, 145, 146; animal studies, 131, 144, 145–147. *See also* Severe combined immune deficiency
Adrenal hyperplasia, congenital, 61, 73; treatment, 74
Adrenoleukodystrophy, treatment, 86
AFP. *See* Alpha-fetoprotein
Agammaglobulinemia, treatment, 75
Albinism, 33; animal analogues, 127
Alpha₁-antitrypsin deficiency, 40, 107; diagnosis, 46, 51; treatment, 81, 82; gene location, 100, 101
Alpha-fetoprotein, in prenatal diagnosis, 53–56
Alpha thalassemia. *See* Thalassemia
Alport syndrome, treatment, 8
Amino acid, 24
Amino acid disorders, newborn

screening for, 60. *See also specific disorder*
Amniocentesis, 43, 54
Amyloidosis, treatment, 83
Anderson, W. French, 93, 126, 180
Anderson disease, treatment, 82
Anencephaly, 6; diagnosis, 51
Animal models, 126–134, 144–147, 193
Autosomal dominant disorder, 20–22, 23–24; detection of, 61–62
Autosomal recessive disorder, 19–20, 22
Autosome, 19

Bacteriophage, 99–100
Base (in nucleic acid), 24; structure, 26
Base pair, 26
Beta globin, 31; gene, 49, 50. *See also* Hemoglobin; Thalassemia
Beta thalassemia. *See* Thalassemia
Bickel, Horst, 59, 69
Biomedical Ethics Advisory Committee, 182
Biomedical Ethics Board, 182–183
Biotechnology Science Coordinating Committee, 181
Biotinidase deficiency, 60